IMAGES
of America

AROUND SHABBONA

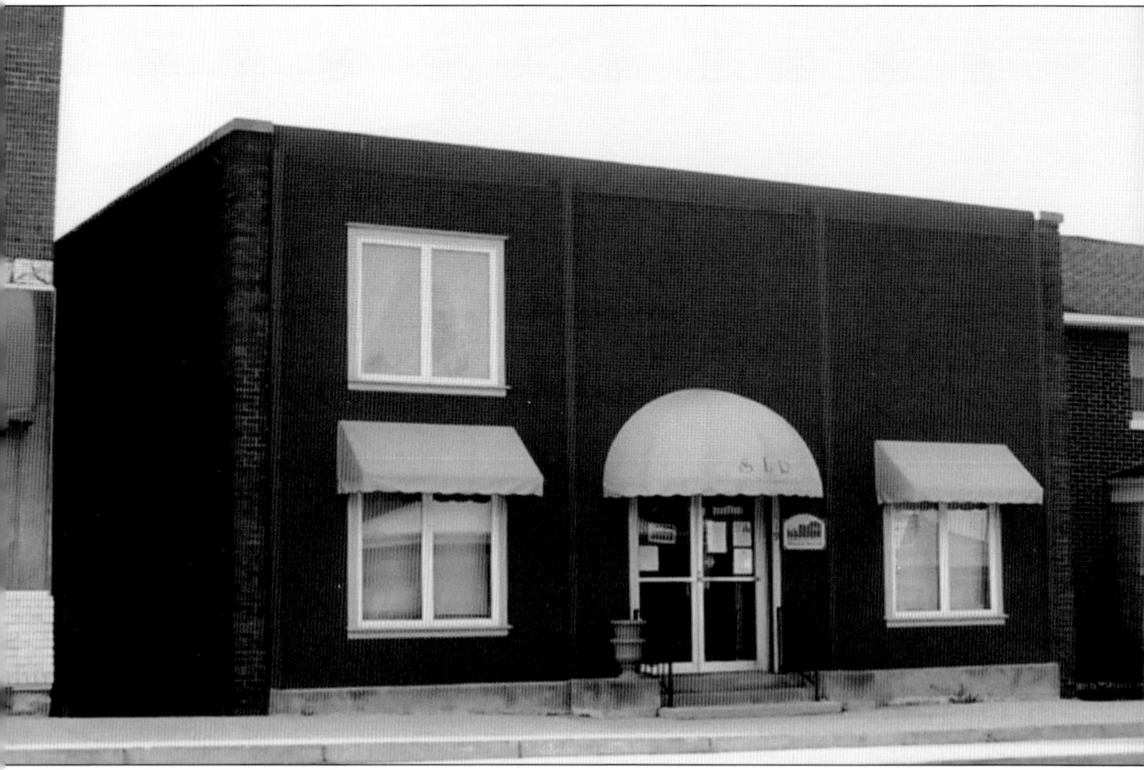

Shabbona-Lee-Rollo Historical Museum opened its doors in September 1992 and is located in the old bank building at 119 West Comanche Street, Shabbona, Illinois. The mission of the museum is to provide a link to the past for any persons with a connection to the District No. 425 school community by enlightening and educating through the extensive collection of historical information. (Courtesy of the Shabbona-Lee-Rollo Historical Museum.)

On the Cover: Allen Park was located south of Shabbona and was a favorite entertainment venue for residents in the early 1900s. Notice the gaslight, flag, and ladies at the food stand in the background. (Courtesy of the Shabbona-Lee-Rollo Historical Museum.)

IMAGES
of America

AROUND SHABBONA

Nancy Fleming Pardridge

ARCADIA
PUBLISHING

ISBN 978-0-7385-8867-4
Published by Arcadia Publishing
Charleston, South Carolina

Printed in the United States of America

Library of Congress Control Number: 2011943025

For all general information, please contact Arcadia Publishing:
Telephone 843-853-2070
Fax 843-853-0044
E-mail sales@arcadiapublishing.com
For customer service and orders:
Toll-Free 1-888-313-2665

Visit us on the Internet at www.arcadiapublishing.com

To all the volunteers over the past 20 years who had an idea for establishing a museum and making that dream a reality

Contents

ACKNOWLEDGMENTS

The Shabbona-Lee-Rollo Historical Museum became a reality on Homestead Day in Shabbona in 1992. Cutting the ribbon were Hazel Clapsaddle, benefactor; Shabbona mayor Fritz Vandre, board member; and Judy Svendsen, Rollo board member. The main exhibit was about the 44 one-room schools that dotted the Shabbona-Lee-Rollo area between 1875 and 1950. Lois Purdy, president of the board of directors, delivered the dedication, and the museum was officially opened. I thank my family for helping me get the many recent pictures of the area; the museum for the information collected throughout the years; Lois Purdy for getting an idea and making it a reality; Mary DeBolt, the first curator; Carol Bend, a former curator whose knowledge about the history of the area and its people were an indispensable asset; Carol and Alice Lintereur for the many hours of helping me acquire and put the information together; Ada Gallagher, the present curator; and members of the Shabbona-Lee-Rollo (SLR) Museum Board for their support and knowledge of the area. Special thanks go to acquisitions editor Chris Metcalf, who gave me encouragement and advice.

Mayor Claudia Hicks and the Shabbona Village Board have provided the old bank building for the museum, and the present Shabbona-Lee-Rollo Historical Museum Board of Directors has been very supportive of updating and maintaining the premises. The board continues to set new goals and plans, as we work to fulfill the plans developed in the past. The Shabbona-Lee-Rollo Historical Museum Board, our curator, and volunteers give many hours of their time to projects and plan to be open for years to come.

Unless otherwise noted, the images appearing in this book were provided by the Shabbona-Lee-Rollo Historical Museum.

Introduction

The contents of this book will include the area covered by the old District No. 425 Shabbona school community to include the towns of Shabbona, Shabbona Grove, Lee, Scarboro, Rollo, and the surrounding area. This area lies in DeKalb and Lee Counties and a portion of LaSalle County. The villages, shown through vintage photographs from the collection at the local museum, share a wide diverse history, from Native Americans like the Pottawatomie to the early settlers who came and persevered during the tough times.

A small amount of the history here involves Chief Shabbona and his family settling in Shabbona Grove, as well as the influence he had on this area. The village of Shabbona and Shabbona Township are both named after the Indian chief. Early settlers began to find their way from the East in the late 1830s, bringing all of their possessions in prairie schooners to make their future homes and playing a big part in the rise of the settlements. This area was attractive because of the excellent land and quality and quantity of timber in the grove. Water was available from Big Indian Creek, which enters the township on section 31, running in a northeasterly direction before taking a southeasterly turn into Paw Paw Township. The settlers were able to build log houses and begin clearing land and tilling the soil. The first settlers were Edmond Towne and Davis Smith, who arrived here in December 1835. Before the building of the railroad, the "Chicago Road," which crosses the township from east to west, became important for eager land seekers and speculators, with the heavy freight wagons following in succession. The towns sprang up because of the railroad. A railway station meant a new village, and a new village meant business and real estate opportunities.

In the best days, Shabbona Grove had three general stores, a tin shop, two hotels, two churches, a Masonic lodge, elevator, and a boot-and-shoe shop. A large business was transacted, and the town experienced steady growth in the early years. But being late in securing a railroad, things came to a standstill. Only a few homes, including a remodeled schoolhouse, remain. Early history of the Ross Grove community south of Shabbona Grove indicates a settlement in 1835. The cemetery still has some tombstones remaining, and they are included in this book.

An English settlement in the western part of Shabbona Township was begun in the fall of 1851. Five young Englishmen—Thomas Wright, Septimus Storey, William Cutts, George Glossup, and Joseph Bidam—came together and were the pioneers of the settlement, taking the land from the government. George Glossup built the first house, which served as headquarters for English immigrants until they could construct homes for themselves. A church and a schoolhouse were built, with a cemetery to come later.

In 1872, the village of Shabbona was surveyed and platted on land belonging to Francis E. Hinckley, John Palen, Catherine Palen, John Ray, Elizabeth Ray, Charles Stevens, and Cynthia T. Stevens, about half a mile west of present-day Shabbona, and given the name of Cornton. A temporary depot was constructed. The site was favorable, however, none of the proprietors would donate land to interested railroad men. Therefore, a removal of the village was determined to be, and Cornton was doomed.

The village of Lee lies upon the line between the counties of Lee and DeKalb. It is mostly in Lee County, a portion equal to about 30 acres being in Shabbona Township, DeKalb County. There have been numerous businesses over the years, however, Lee has not experienced any great growth except for in the early years of its existence. The 1880 census showed a population of 218. Today, it is in the low 300s. Also, a small amount of the history of Scarboro in Willow Creek Township, Lee County, will be covered. The Norwegians who settled here began immigrating to the area in the early 1800s and brought their traditions with them. Land was cheap, and the thick groves of oak trees and area along waterways provided good homesites.

Present-day Rollo has three homes, Congregational Church, a parsonage, Mullins Grain Elevator, Paw Paw Township Building, a park, and a one-room schoolhouse. According to the *DeKalb Chronicle Illustrated Souvenir* 1899 edition, Paw Paw Township is noted for being one of the most prosperous agricultural centers in DeKalb County. The township also was known for its advancements in education, like establishing the teacher's seminary. The East Paw Paw Teacher's Institute and Classical Seminary was located west of Rollo at Chicago and County Line Roads, with the bell from the old school hanging today in the steeple of the Congregational Church in Rollo. There is a burial site in Stevens Cemetery, southeast of Rollo, for Katherine Van Tassel Lake, the Katrina that Washington Irving writes about in "The Legend of Sleepy Hollow."

This pictorial journey will include photographs from the past and present. Some of the highlights included are the Shabbona Lake State Park, Shabbona Forest Preserve, and Shabbona Country Club, which bring many visitors to our area. Several centennial farms in the area are also previewed. One-room schoolhouses, some having been made into homes for present-day use, are also covered in this book. There was much excitement in the 1930s over the fact that the possibility existed of oil beneath many of the farms in the area. Also during World War II, the government needed a supply of fiber for cordage for all types of war needs. A plant was built north of Shabbona, where as many as 100 people were employed. Many displaced Japanese Americans were housed on-site for a few years. German prisoners from Camp Grant were brought to the area to work in the fields.

Almost all of the remarkable photographs in this book have been given to the museum over the 20 years of its existence. Without these people opening up their collections and the many volunteers who organized the material, this book would not have been possible.

One

Early Years

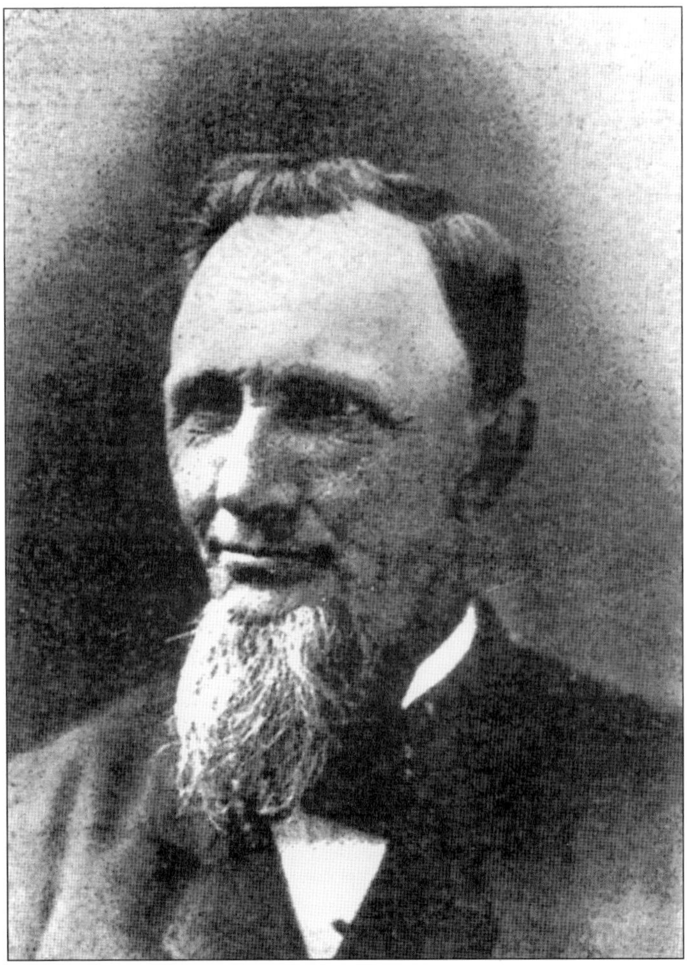

The settlers began finding their way from the East in the 1830s, bringing all their possessions in prairie schooners. Peter V. Quilhot, a farmer and a pioneer of Shabbona Township, had an improved farm of 170 acres. He was born in New York on January 21, 1822. He lost his father as a youth, and circumstances rendered it necessary for him to aid his mother in the care of the family.

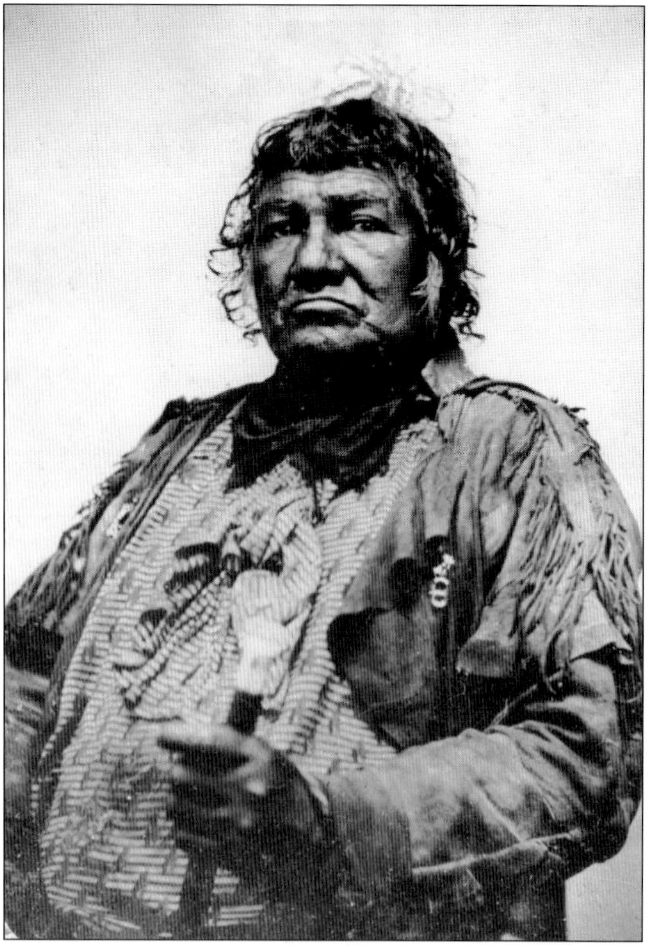

Shabbona or "Built like a Bear" was born at the village of the Ottawas in Canada in 1775. In height, he was five feet, eight inches; broad in the shoulders and chest; and weighed about 200 pounds. Shabbona died in Morris, Illinois, on July 17, 1859. In 1903, a granite boulder was erected as a monument on his grave site in Evergreen Cemetery. (Courtesy of the Chicago Historical Society.)

Pokanoka, daughter of Chief Spotka, was the third wife of Chief Shabbona. She was a large woman, weighing over 300 pounds. Shabbona's second wife, Nebebaqua, and Pokanoka, mother of his four sons and four daughters, lived side-by-side but not harmoniously. Pokanoka, who drowned in about six inches of water, died in 1864. She is buried beside her husband. Pokanoka is standing on the far right in the picture.

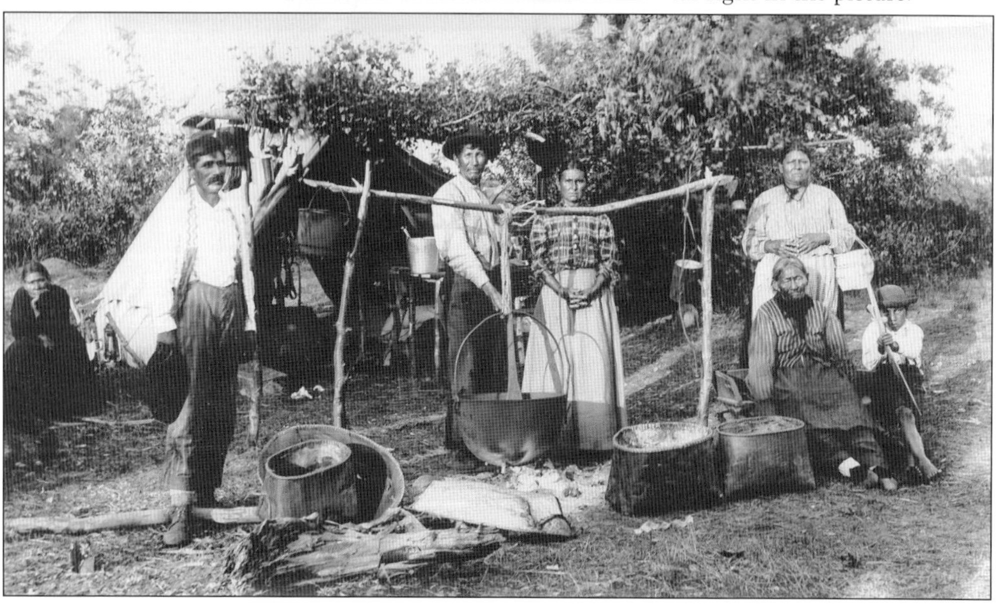

As a part of the hunters from the Lake Erie region, Chief Shabbona of the Ottawa Tribe made his first visit to Illinois in the fall of 1800. The hunters were quite at home among their friends, the Pottawatomie. When they returned to their homes, Shabbona remained for the winter at the home of Spotka, the chief of the Pottawatomie. Chief Shabbona married Pokanoka of the Pottawatomie, becoming a member of their tribe and, eventually, the chief. They lived near Ottawa on the Illinois River. Later, they made their home near Shabbona Grove, and their band then numbered about 139. The memorial sign was erected by the state of Illinois at the corner of Highway 30 and Indian Road in 1934. It is now in the museum. In the picture below is Chief Shabbona's daughter Martha, seen at center, in Ottawa in 1898.

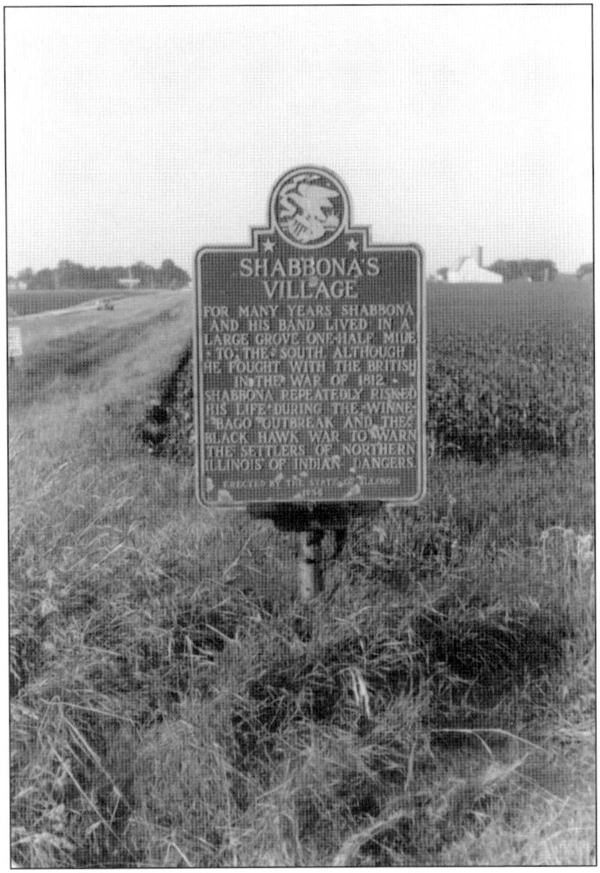

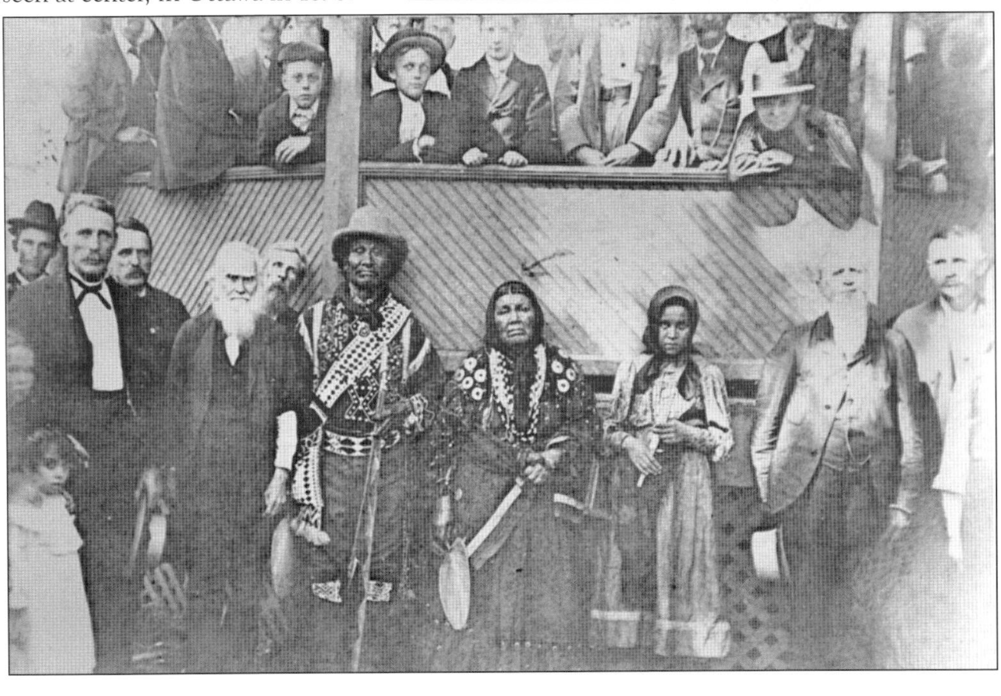

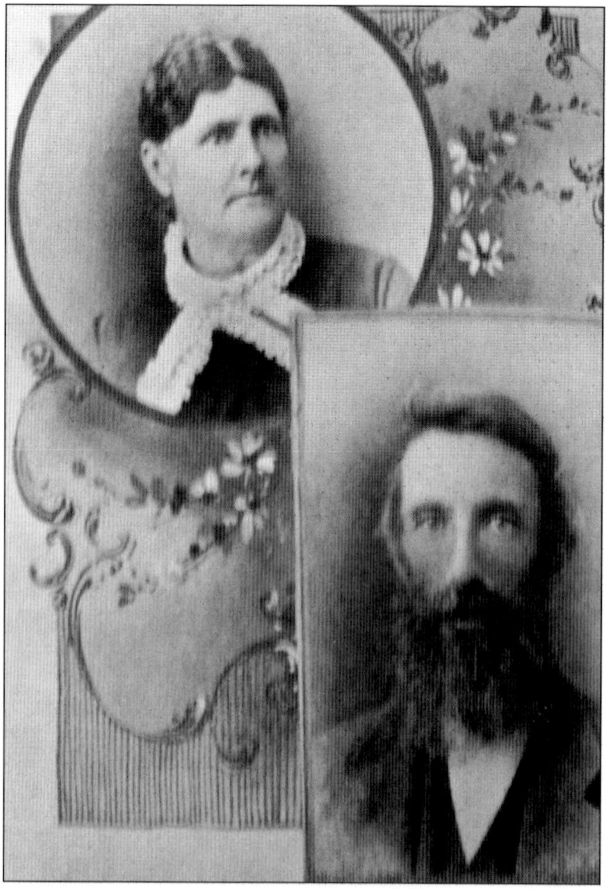

David and Lucinda Norton lived in a log house across the road from the entrance to the present-day Shabbona Forest Preserve. Chief Shabbona would often visit the Nortons after hunting. Mrs. Norton stated, "After hunting, the Chief would bring the venison to my mother. When he came he always rapped on the window which frightened me, but I felt safe when he was near." In 1837, Shabbona was notified by the Indian agent that, by terms of the late treaty, "All members of his band, with the exception of those of his own family, must remove to their new reservation in Western Missouri." In 1849, during Shabbona's absence, the commissioners decided that he had forfeited his right to his land by moving away and that the land should be sold. Below is a postcard view of early Shabbona around 1900.

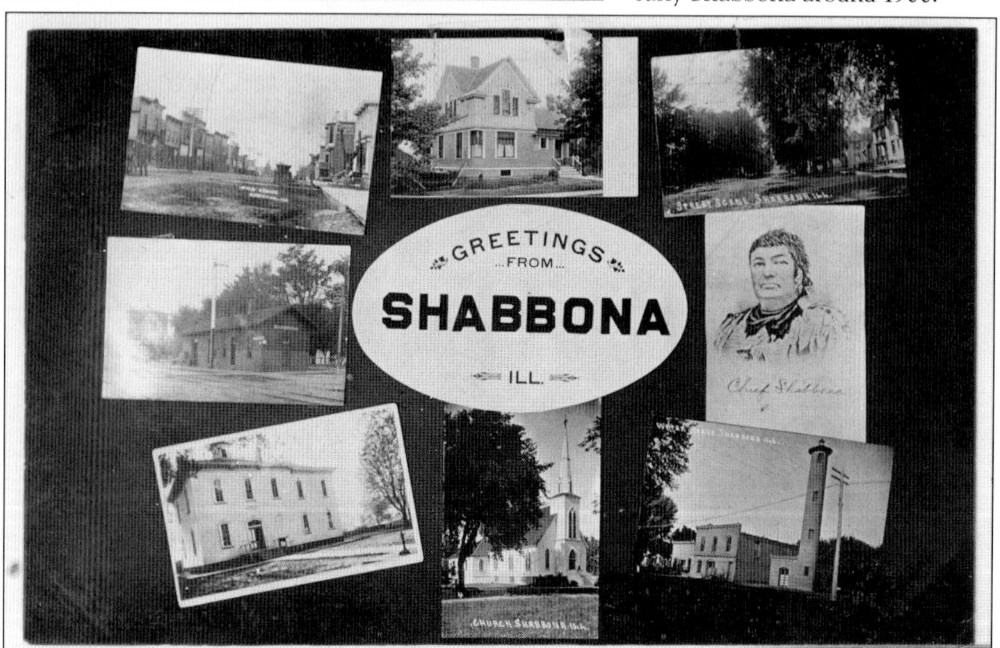

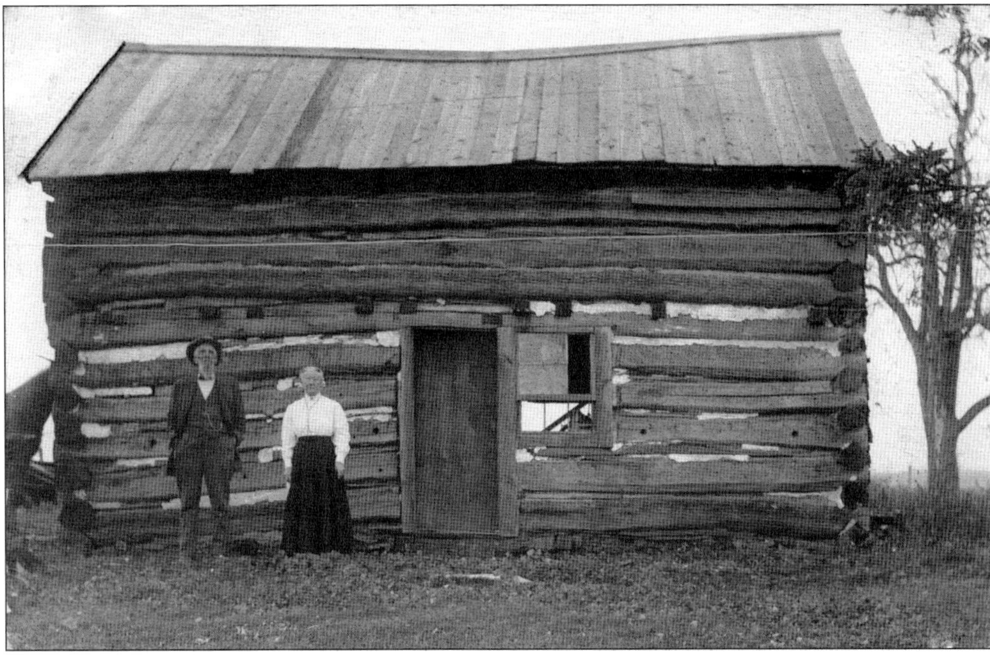

Henry Husk came to Illinois from New York when he was 13 years of age. He and his brother William walked from Chicago to Shabbona in two and a half days, arriving at the home of their aunt Elizabeth Quilhot Miller, where they made their home. Henry Husk drove a stagecoach from Shabbona Grove to Galesburg and also did some farming. He was deputy sheriff of the county and was considered one of its most efficient officers. In later years, Henry Husk was a personal friend of Chief Shabbona and would often be his guest at the Indian village near the Husk home. Henry Husk died at his home in Shabbona on March 23, 1927. Henry Clapsaddle, shown with his two grandsons, was a pioneer in Paw Paw Township. He came in 1852 and was active in community affairs.

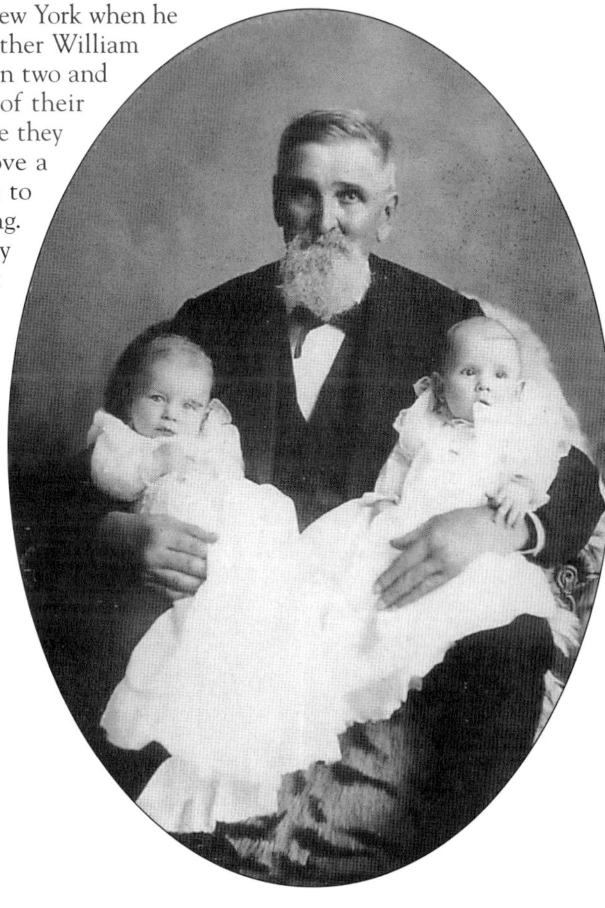

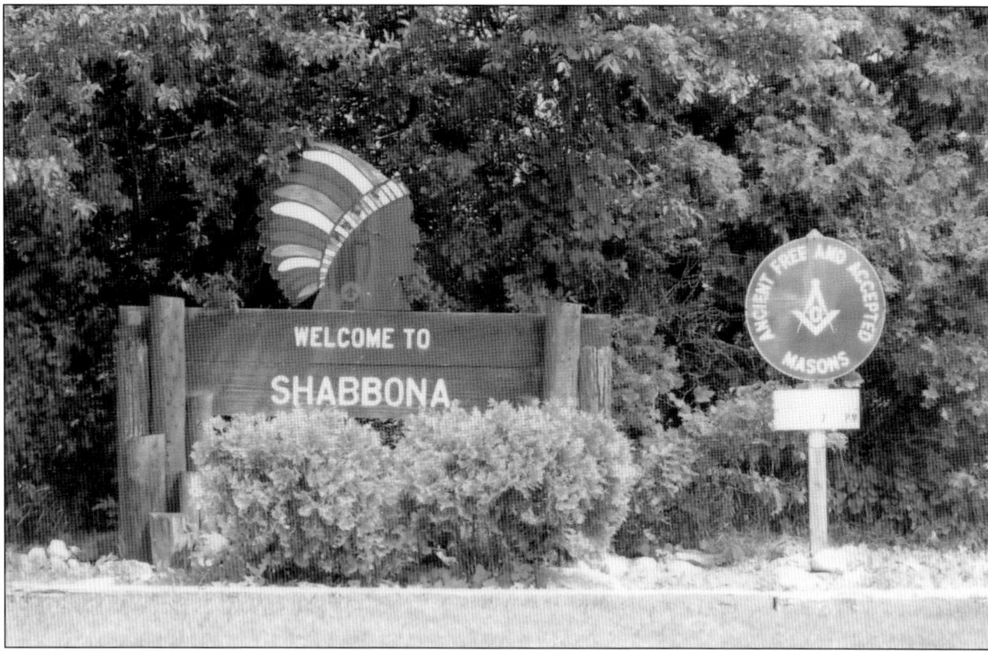

In 1855, just southwest of today's Shabbona site, there was a post office called Malma in a farmhouse on Maryott Hill. The mail was brought by stage to Shabbona Grove and to Malma. Ira Fisher was its first postmaster and was succeeded by John Maryott. In 1870, the Chicago & Iowa Railroad was built, and a small village named Cornton sprang up farther west of where the village of Shabbona is now located. However, a depot was built farther east on a donated piece of land. In the spring of 1872, the town now known as Shabbona (named after the famous Indian chief Shabbona and the township) was surveyed and platted out. W.H. Ray erected the first building in the village in the fall of 1872 and was the first station agent and postmaster. (Below, courtesy of the Hallam Family Collection.)

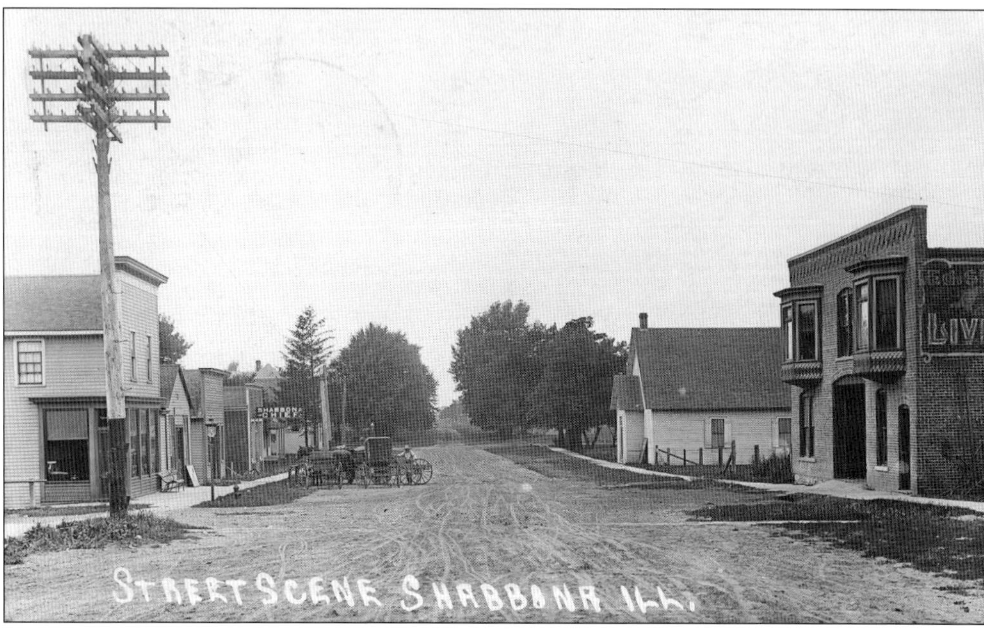

In the spring of 1845, John Palm moved with his family to Shabbona Grove from Ohio and entered 160 acres of government land on Section 14. Because of bad vision, he eventually rented his farm and moved into the town of Shabbona. He was elected as a member of the first board of village trustees. It was largely due to his liberality and public enterprise that the railway station was established here. He generously gave the undivided one-half interest in 40 acres of land for a village plat and also gave a number of lots to encourage the growth and settlement of the village. He is credited as one of the founders of the village of Shabbona. Below is an early photograph of Shabbona. Notice the garage and harness shop on the corner of Illini and Comanche Streets. (Right, courtesy of Becky Krone Knisely; below, courtesy of the Hallam Family Collection.)

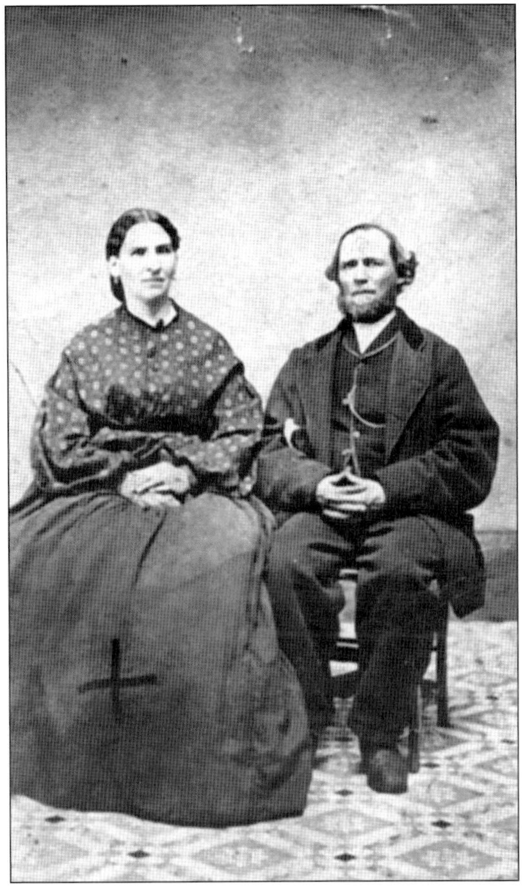

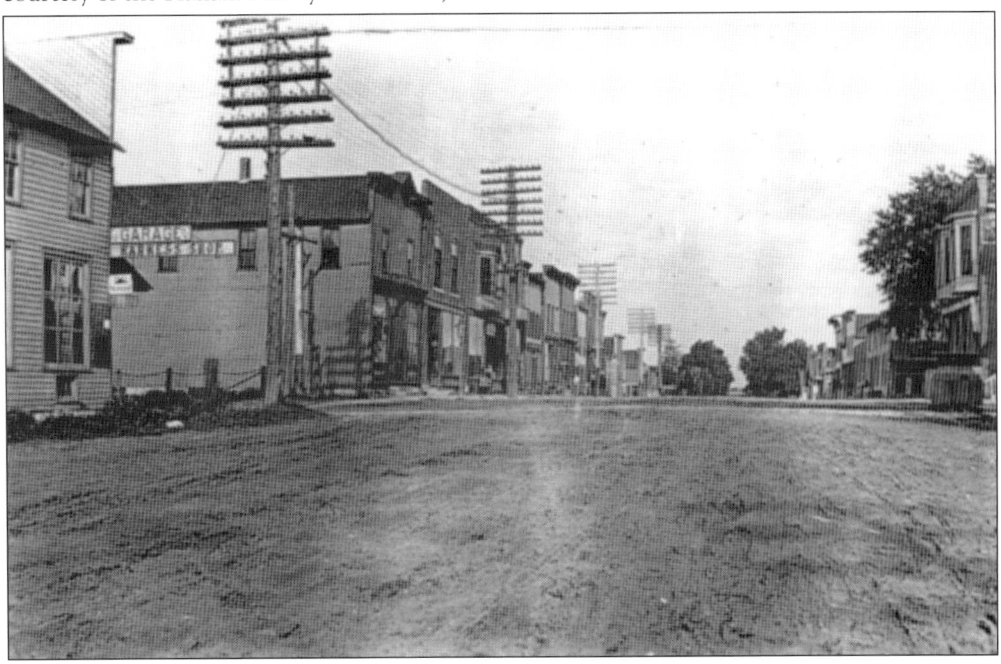

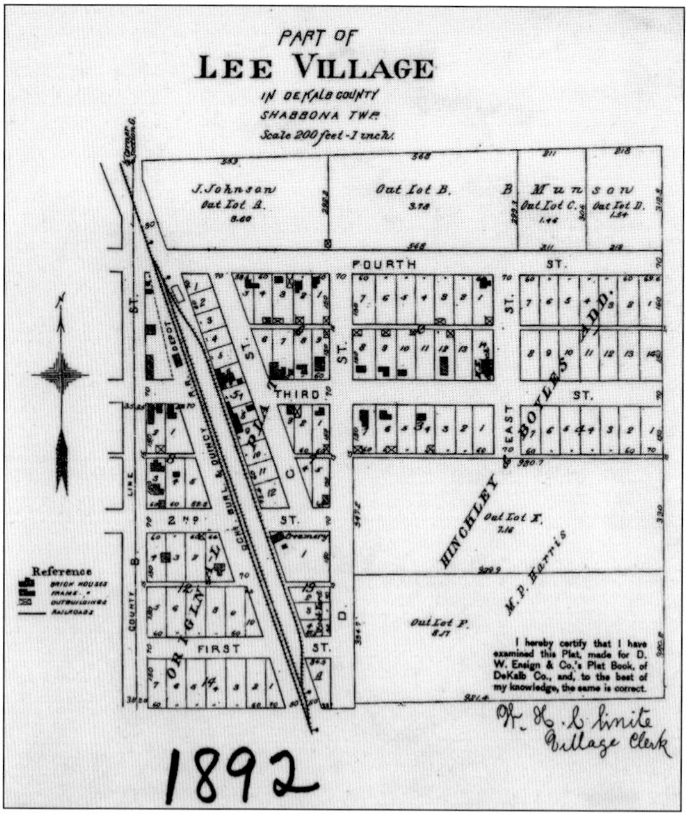

The US government granted patents for tracts of land in the vicinity of what is now Lee, Illinois, to Hugh Boyle, John S. Miller, and John Kennedy. In 1871, the big news was that the Chicago & Iowa Railway would install a switch track and depot and provide regular rail service. Shown is the 1892 plat of the DeKalb County side of Lee. (Courtesy of LuAnn Benson.)

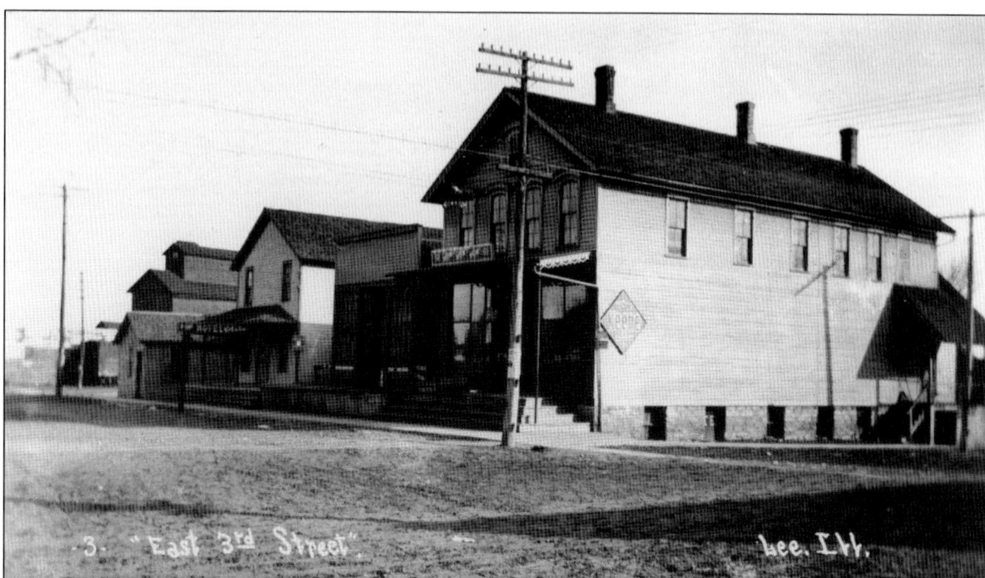

"Lee Station" was the first name for the settlement, as it was the first point of entry in Lee County by the railroad. By 1872, the village asked for corporation, and both counties granted it in 1874. Lee was named after Richard Henry Lee "Light Horse Harry," the Revolutionary War hero. The picture shows East Third Street on the DeKalb County side of early Lee, Illinois.

Rollo is a small village situated on the Chicago & Northwestern Railroad, seven miles northeast of Earlville, in DeKalb County. About the time the little village was settled, a series of children's books featured Little Rollo Holiday (created by Jacob Abbott in 1835) as a popular character. According to the *Earlville Leader* on March 19, 1903, Rollo is the capital of Paw Paw Township, DeKalb County, Illinois.

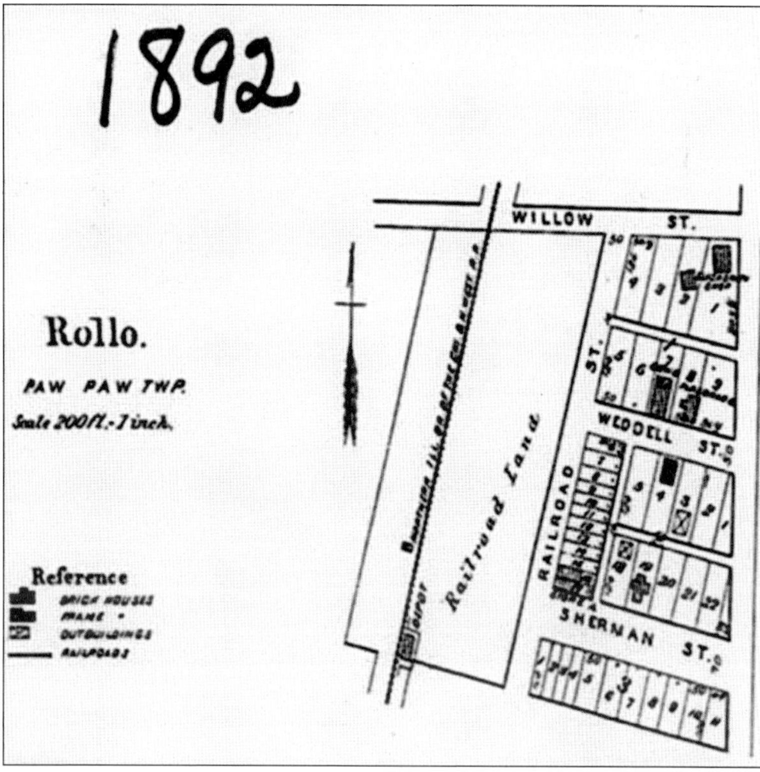

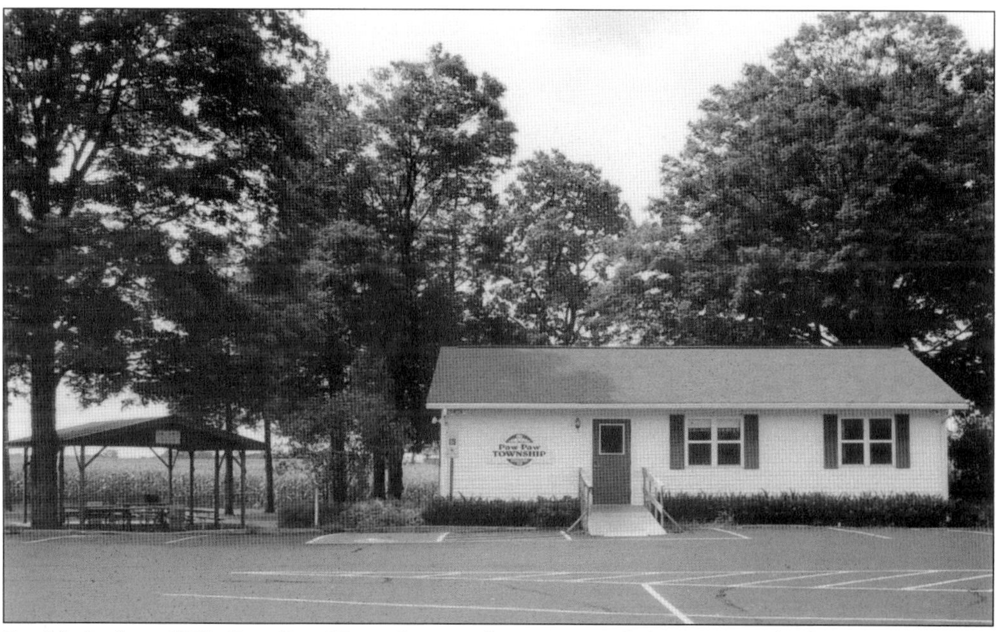

Established in 1850, Paw Paw Township residents enjoyed three natural advantages: several scattered groves of timber, numerous streams, and rich prairie soil. The name Paw Paw comes from the tree of the same name, which is known for its peculiar and juicy fruit. The Paw Paw Township Building is located on the old Rollo School site. Lake Family Shelter, donated by the Dean Lake family, was built next to it.

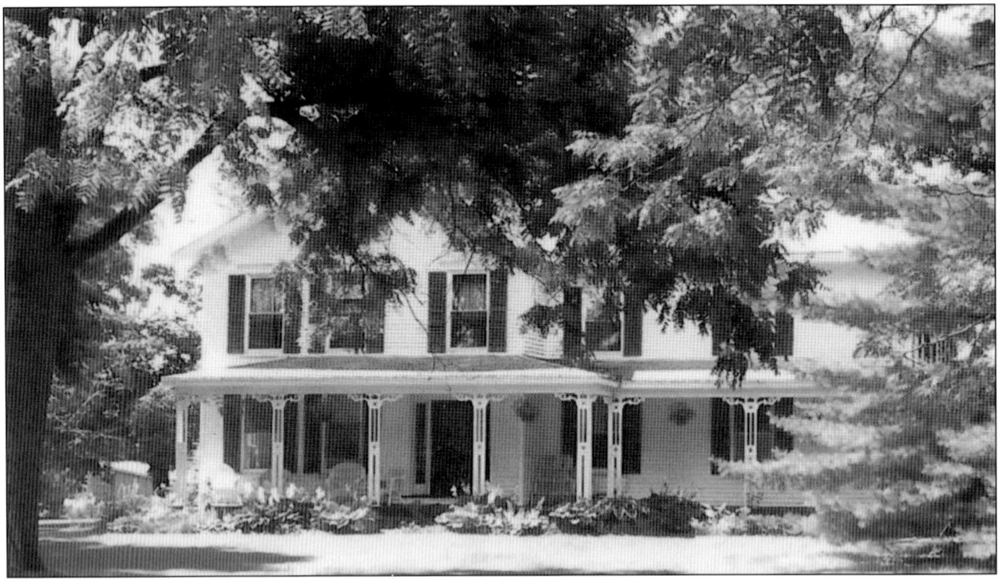

Often referred to by early settlers as landmarks, the four groves in Willow Creek Township are Smith's Grove, Allen's Grove, and North and South Twin Groves. The Indian trail from Ottawa north to the hunting grounds in Wisconsin ran through Allen's Grove. Brothers Lewis H. and Gilbert Durin came from Vermont in 1849 and left an imprint on Willow Creek Township. The station "Scarboro" on the Rochelle Southern Railway was located on L.G. Durin's farm. Sunny Lawn is the home of Steve and Diane Durin in Scarboro, Illinois. It was built in 1869 for L.G. Durin and is a rambling frame house with a big front porch. The log cabin south of the creek was used for the kitchen. The bell below sits in the center of Scarboro as a remembrance of the Scarboro Community Church.

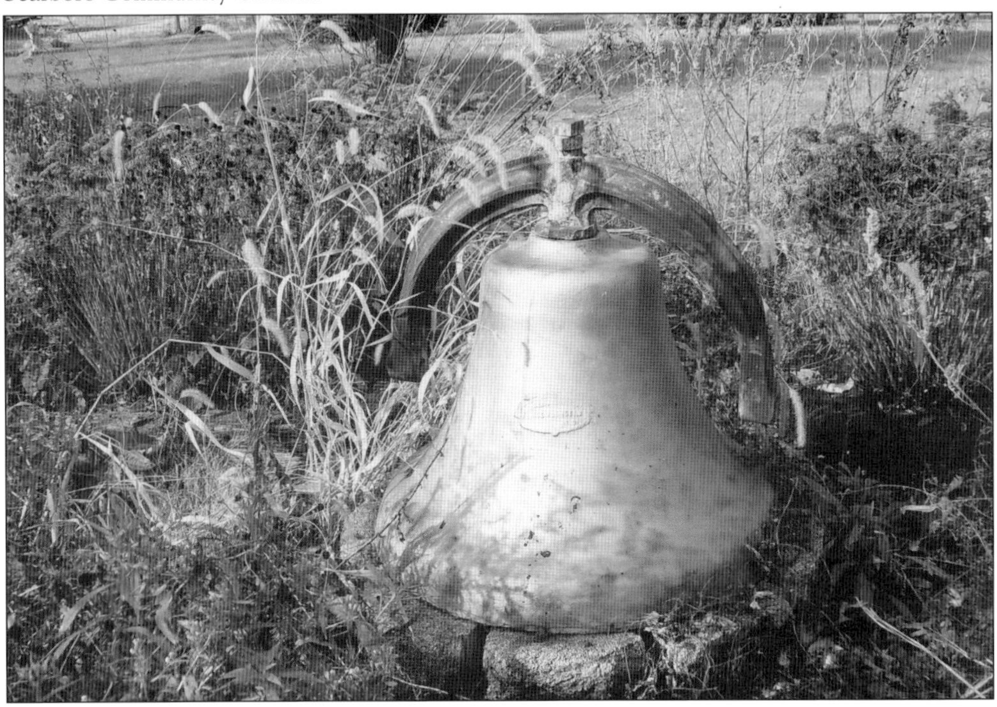

Two

One-room Schoolhouses

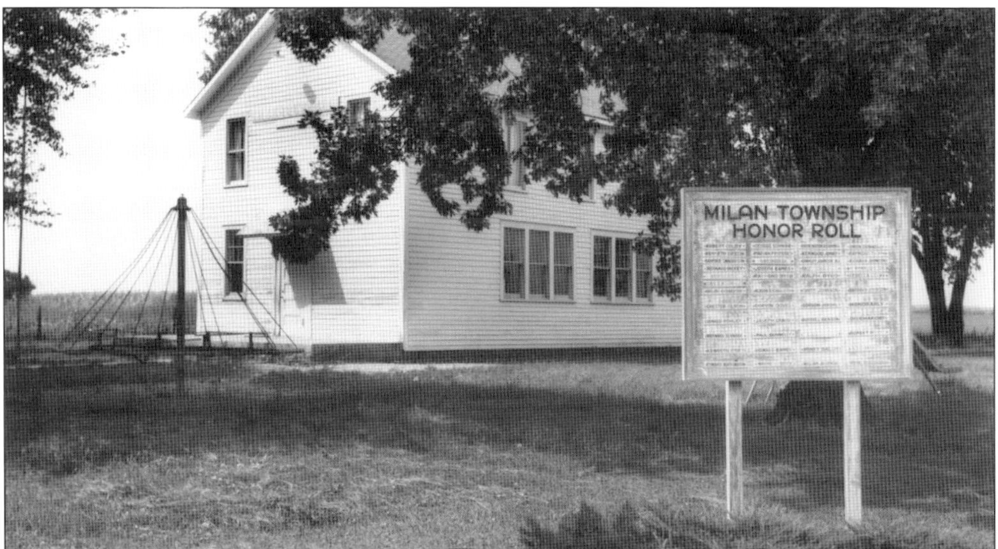

One-room schools, like the Milan School Center, were scattered throughout the area. Each school had a director whose job was to hire the teacher and make sure there was coal for the furnace. Buildings were heated with coal stoves, had outhouses, swings, and a ball diamond. Water was obtained from a well or hauled from a neighbor's. These schools were closed in the 1940s, and children went to public schools.

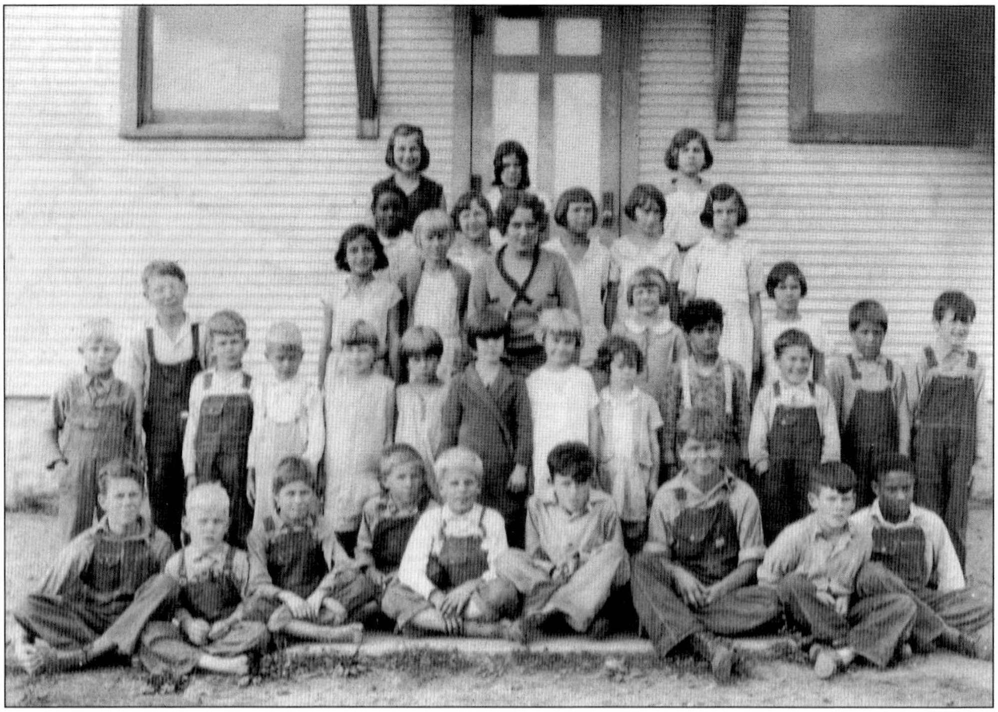

Shabbona Grove School District No. 1 listed William Curtis as the first schoolteacher of record in Shabbona Township. School was conducted in the home of W.C. Olmstead in 1842. The first schoolhouse was built in 1843, west of Indian Creek on the north side of the road. It burned, and a new schoolhouse was built in 1846 and moved to its present site in 1868. The school picture above is of the 1931 or 1932 class, with teacher Pearl Frost. The school closed in 1945 and became the private home of Frank and Nellie Whitehouse for many years. Below is a photograph of Kay, a member of the Hallam family, who looks as if she is ready for school, posing with her two pals, a cat and a dog, and the Cruso Express wagon. (Below, courtesy of the Hallam Family Collection.)

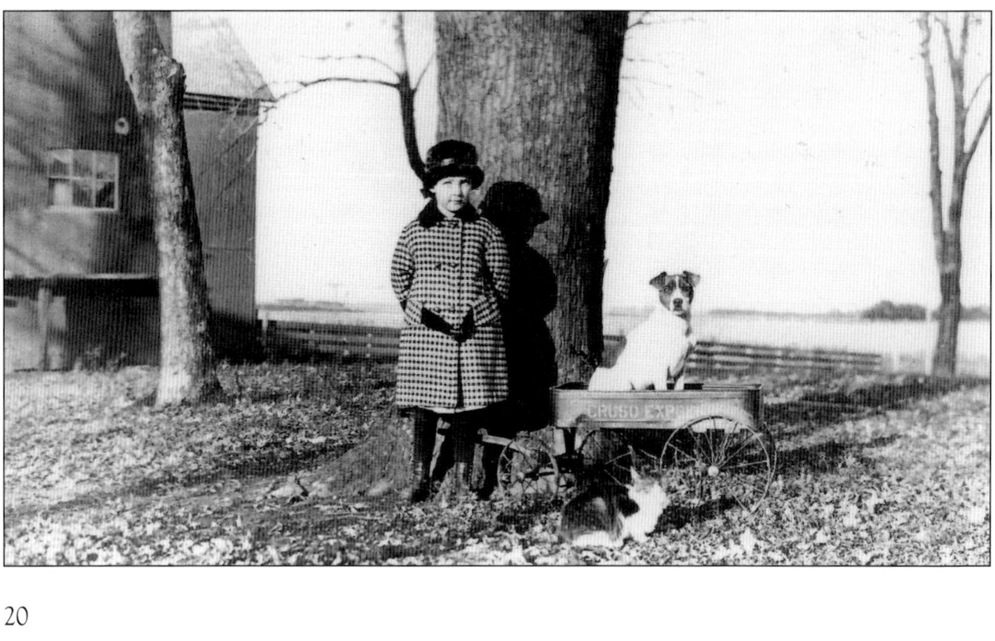

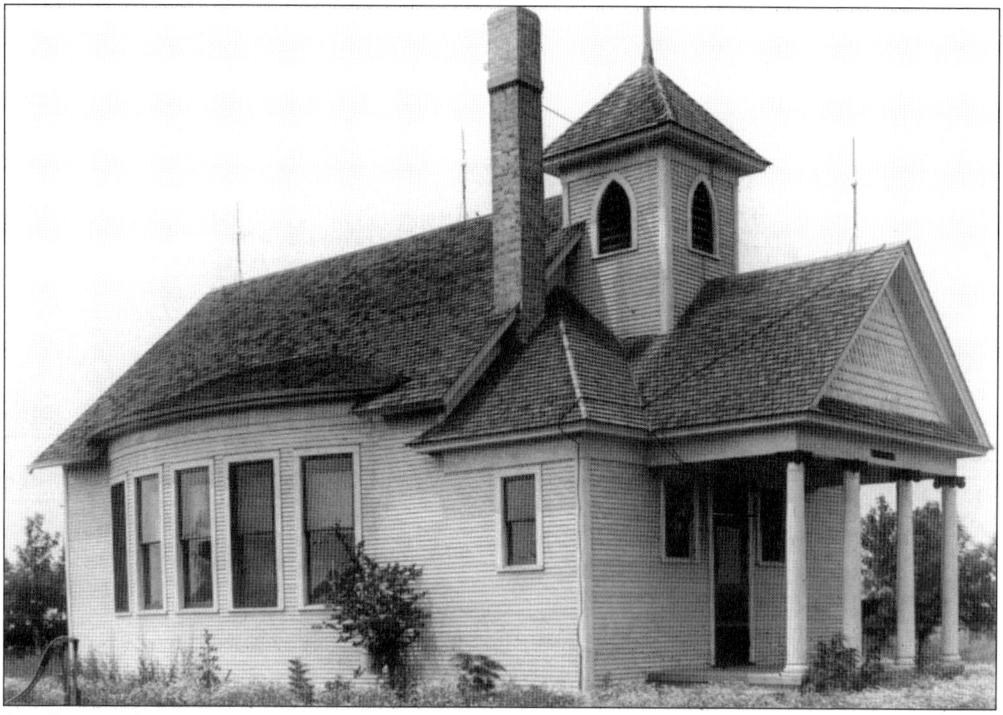

Scarboro and Twin Grove School District No. 140 was a log cabin erected in 1852. A new school was built in 1911 for a total of $2,200. The following is from an article in the *Dixon Weekly Citizen* on May 22, 1913: "Architecturally, the building is perfect. In sanitary appointments, it was more than perfect. The library is a gem and stocked with books such a school demands." It is now a private home.

Hilleson School District No. 143, located at the corner of Woodlawn Road and Route 30 in Willow Creek Township, closed on July 1, 1948, and one part of it joined District No. 271. In 1949, the remaining part combined with District No. 425. The school had been closed for four years prior to being reopened in 1942 with Orla Nangle as the teacher. When the enrollment reached less than 15, country schools began to close due to legislation.

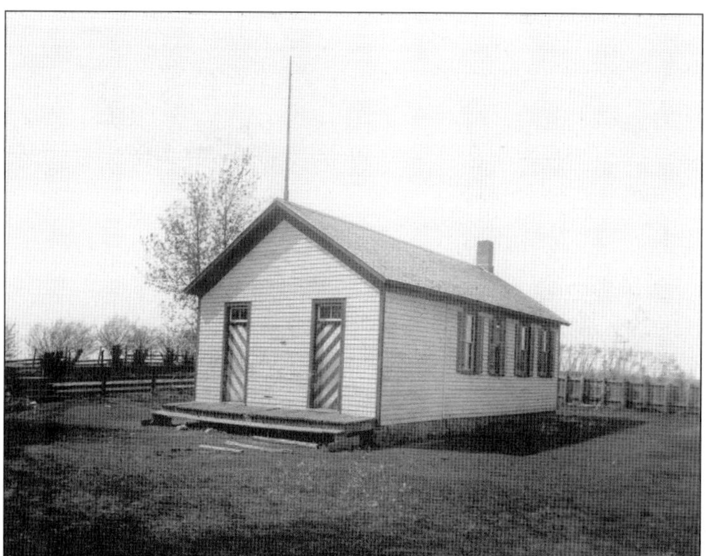

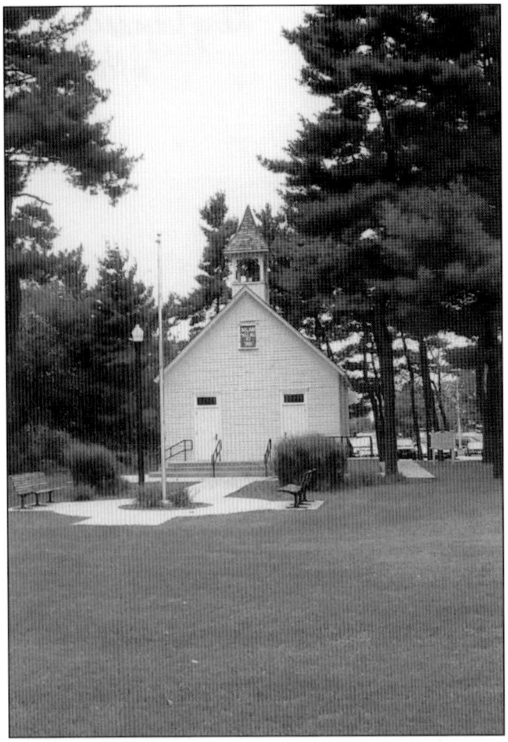

The Tysdale and Berg School District No. 83 schoolhouse was built around 1900 and located on the corner of Tower and Perry Roads in Milan Township. It was closed in 1942 and moved to Northern Illinois University to be dedicated in 1999, becoming part of the Blackwell History of Education Museum. It was notable because it was built with many new conveniences, including a coal-burning furnace in the basement.

The Nowe School District No. 123 building was located on Lee Road about a quarter of a mile east of Tower Road. After the school closed, it was moved to Lee in 1919 and remodeled as the private home of George and Essie Cutts. Scuff marks from benches and burns from the stove still show on the floors. The slogan above the door reads, "Not how much but how well." (Courtesy of Helen Mullins.)

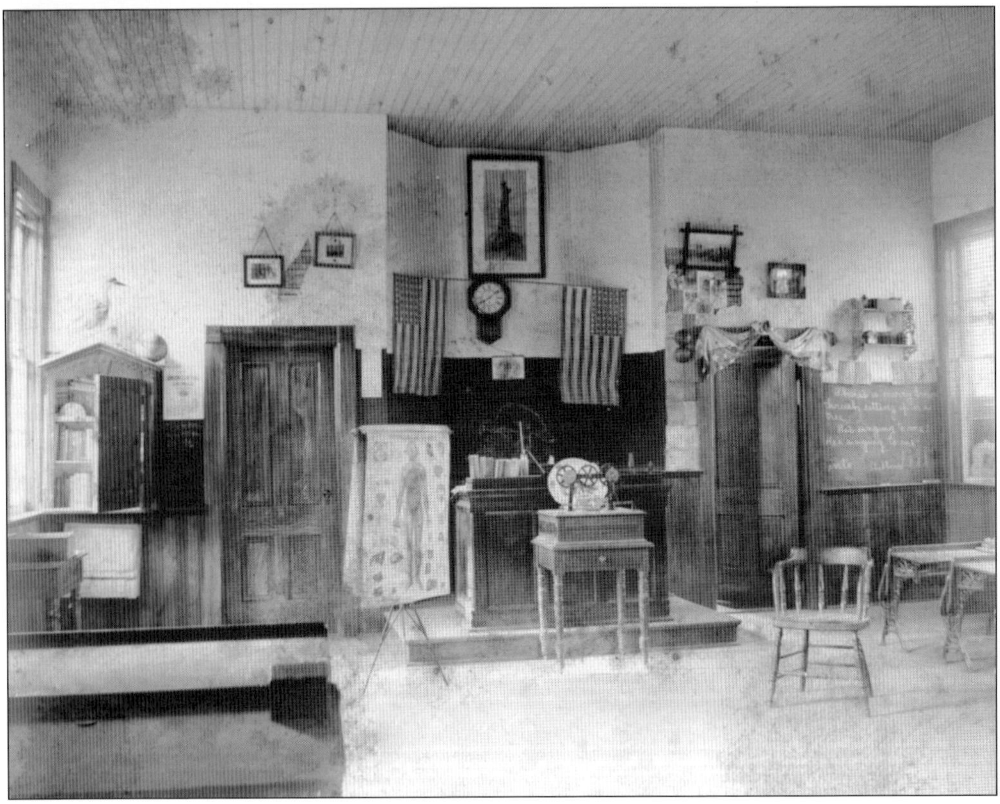

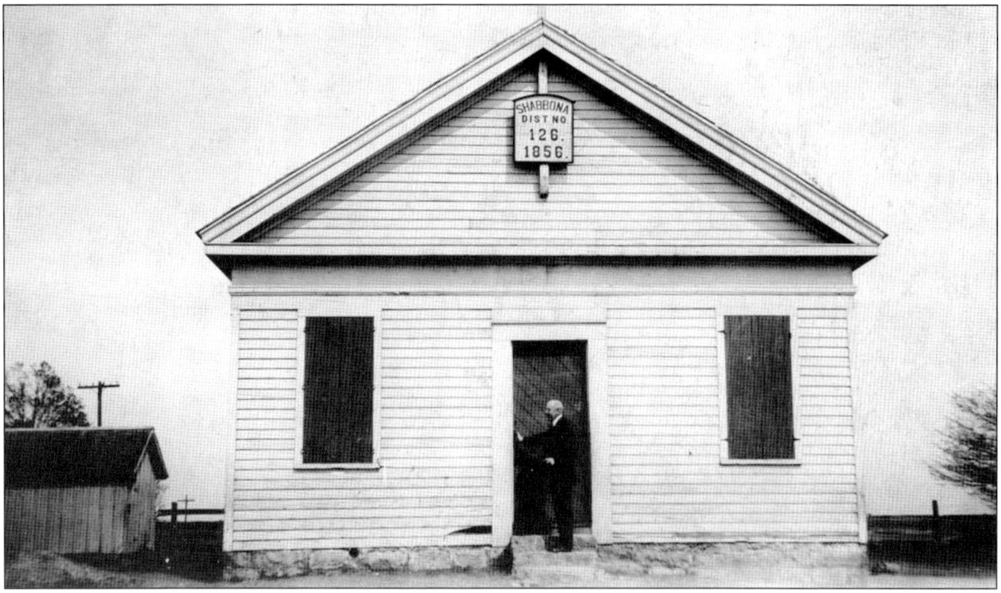

The Quilhot School District No. 126 was built in 1856 at the southeast corner of University and Preserve Roads, located east of Shabbona. Thomas J. Terry donated land. According to the superintendent of schools for DeKalb County, it was a large school with 54 students, 31 recitations, and a very small blackboard. Leona Quilhot taught from 1944 to 1947. It closed in 1948 and became a workshop for Miles Quilhot.

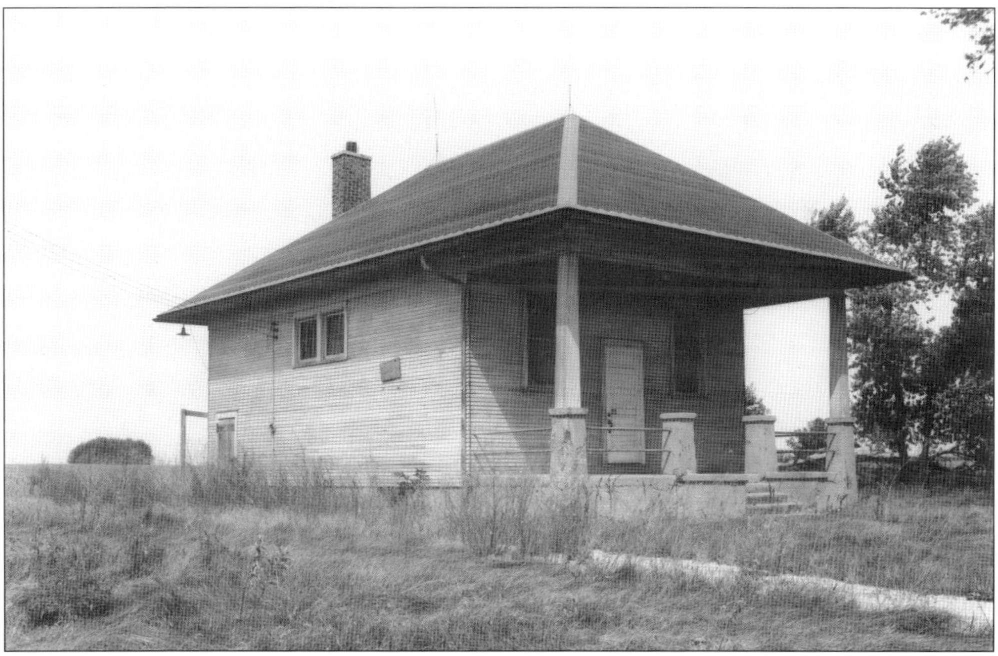

The Wormley School District No. 122 was located at the northwest corner of the intersection of Shabbona and Lee Roads. It had a cement porch across the front, where the children could play in bad weather. It also had a cistern pump in the basement, where the children could wash their hands before lunch. It was built by the Svihus brothers and is now a private home. This photograph was given to the museum by Otta Ritzman.

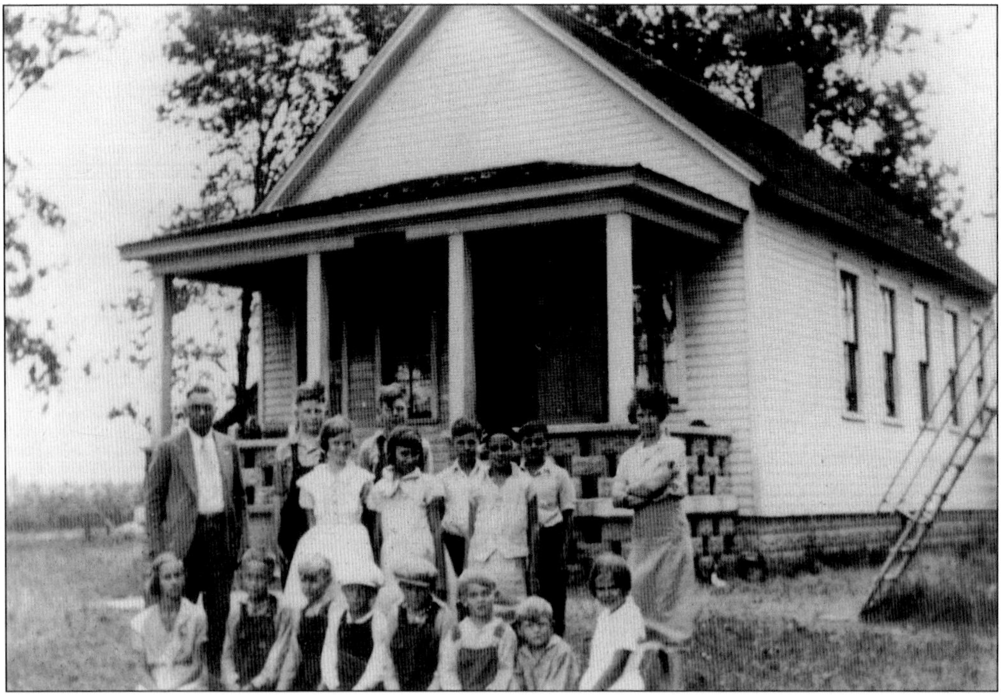

The Peterson School District No. 137, found north of Lee, was located at the southeast corner of Woodlawn and Reynolds Roads in Alto Township. It was built in 1885 and then burned and was rebuilt in 1917, closing for the final time in 1952. Nels and Ragnilda Peterson donated the land. Alice Tysdale and Gordon Beals both taught at this school. The schoolhouse was remodeled into a private home.

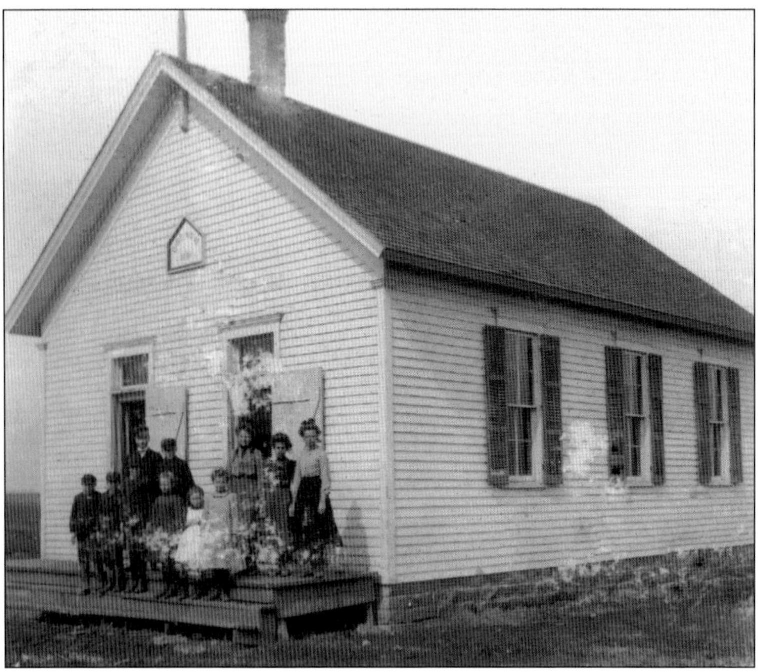

Prestegaard and Bly School District No. 139, located north of Lee at Herman and Woodlawn Roads in Alto Township, was built in 1865. The land came from the O.J. Prestegaard farm. The school closed in the 1940s. Otto Oleson bought the building in the 1950s, and someone set fire to it in 1967. Half of the school became District No. 271, and the remainder of the students went to District No. 425.

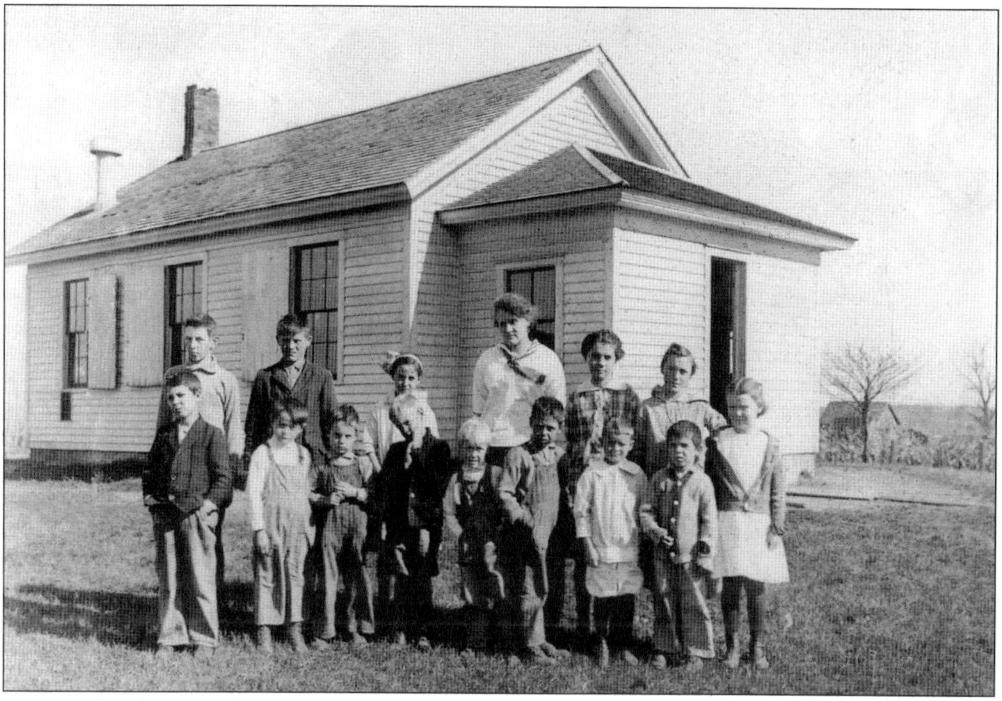

Morse School District began on March 19, 1859. It was located on Shabbona Grove Road, about three-fourths of a mile west of Shabbona Road. Morse sold land to the school for $1. The district was originally Shabbona Township School District No. 5, but the 1905 plat book called it District No. 128. The building burned in 1917. It was rebuilt farther west and opened as Olmstead School. Clarence Hodge was the last teacher.

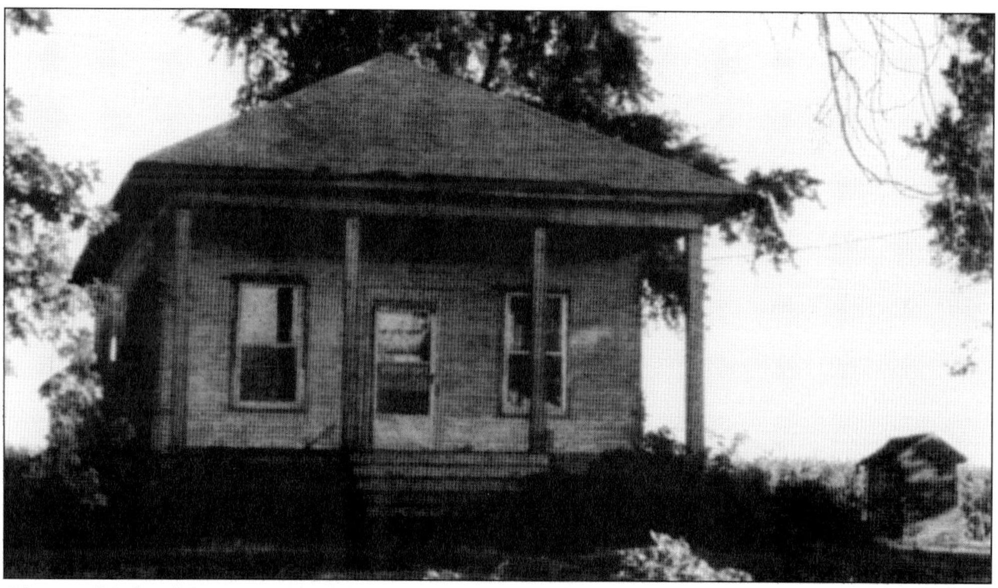

Olmstead School District No. 128 was established in 1918 when Morse School burned. Elizabeth Lovering sold land on the north side of Shabbona Grove Road and west of Shabbona Road to school trustees for $225. May Beiler taught from 1942 to 1945, and Doris Wells Firkins taught for eight years. The school closed in 1950, became a private home, and was later demolished.

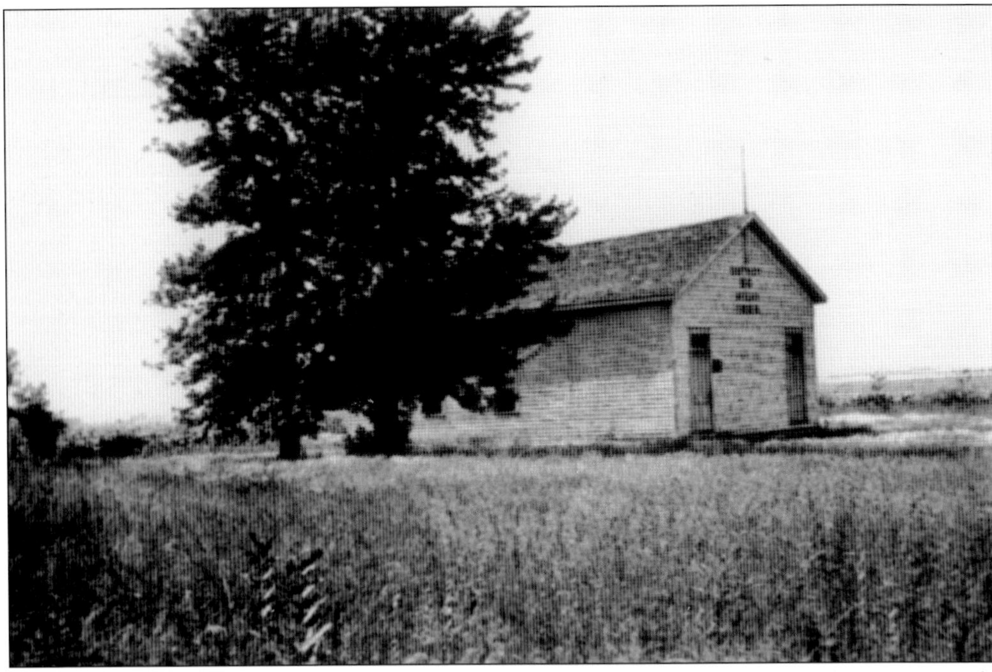

Toad Hollow District No. 84 was located at the northwest corner of Tower and McGirr Roads. The school was erected in 1868 and named after the Toad Hollow area in which it was located. In 1870, there were 55 students enrolled at the school. Thelma Hansen Olson went to school there and also came back to teach after attending Northern Illinois University for two years. The school had a cyclone cellar below the cob and coal house, and water was carried from the Stanley Johnson farm. The school closed in the spring of 1944, and the land reverted back to the Henry Kittleson farm. The building was eventually demolished, and John and Linda Hopwood, daughter of Thelma Olson, built a home on the site. The teacher in this 1910 photograph is Jennie Prestegaard Munson.

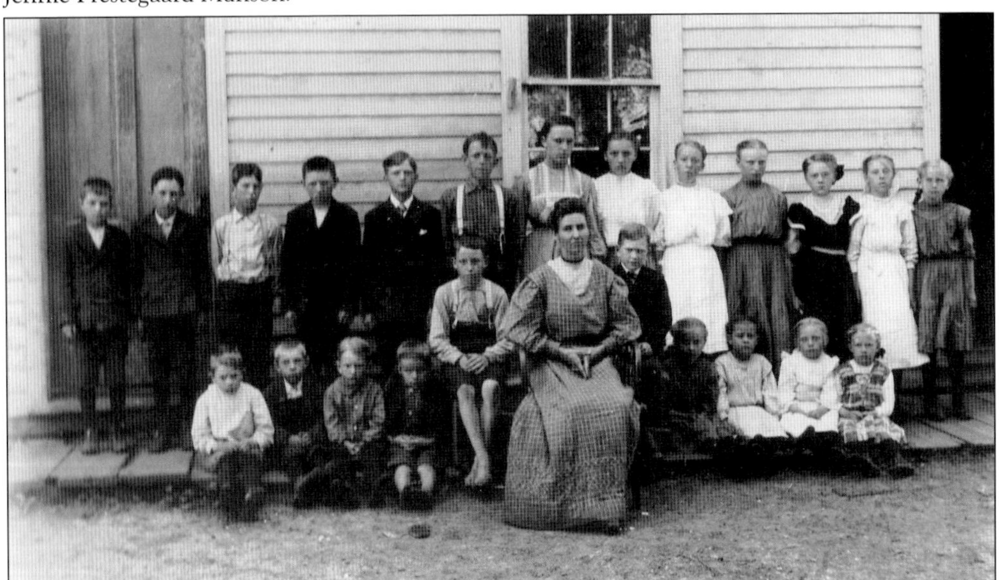

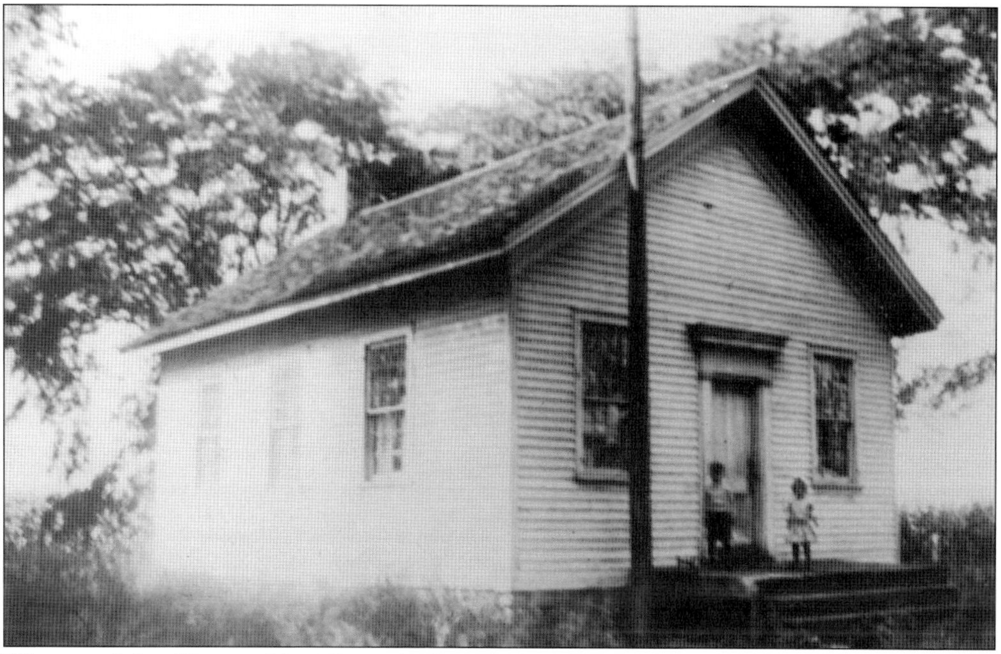

Whitman School District No. 137 was built on the northwest corner of Svendsen and Hyde Roads before 1871. It was named for the Whitman family, who owned the farm across the road from the school. It consolidated with Rollo School in 1912. After the school closed, the building was moved to the Beth Powers Morgan farm and used as a garage.

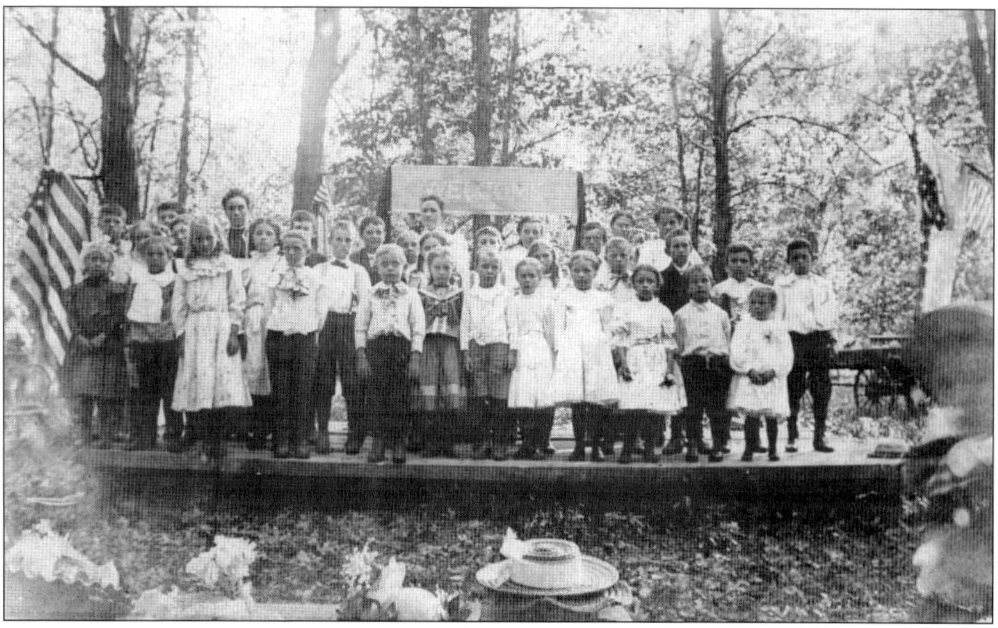

Ross Grove School District No. 131 was started in Paw Paw Township when Benjamin Harris had started a private school in his home in 1836. The first log schoolhouse was constructed in 1845, and the first frame schoolhouse was later built in 1850 on the northwest corner of Chicago and Sleepy Hollow Roads. Ross Grove School, named after Joseph Ross, consolidated with the Rollo Consolidated School in 1912. (Courtesy of Vince Cox.)

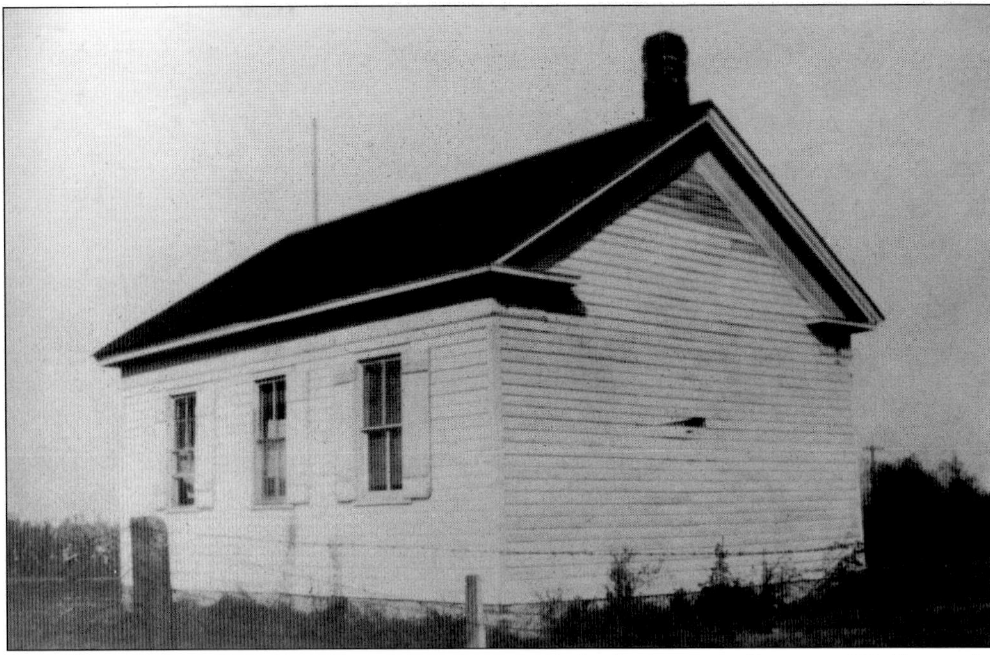

In 1857, Rollo School District No. 134 was originally built about a mile west of Rollo on the northwest corner of West Suydam and South Paw Paw Roads at the Furnas Farm. In about 1886, when the Chicago & Northwestern Railroad went through Rollo, the school was moved to a location near where the Rollo Consolidated School was later built. In 1912, classes were held in this building until the new school was completed. Rollo School District No. 134 consolidated with the Rollo Consolidated School in 1912. The building was moved to its present location near the elevator. It was used as the Paw Paw Township Hall for several years. The building is now owned by the Mullins family and is not being used. The picture below is of the old Rollo High School scoreboard, which is on display in the Shabbona-Lee-Rollo Historical Museum.

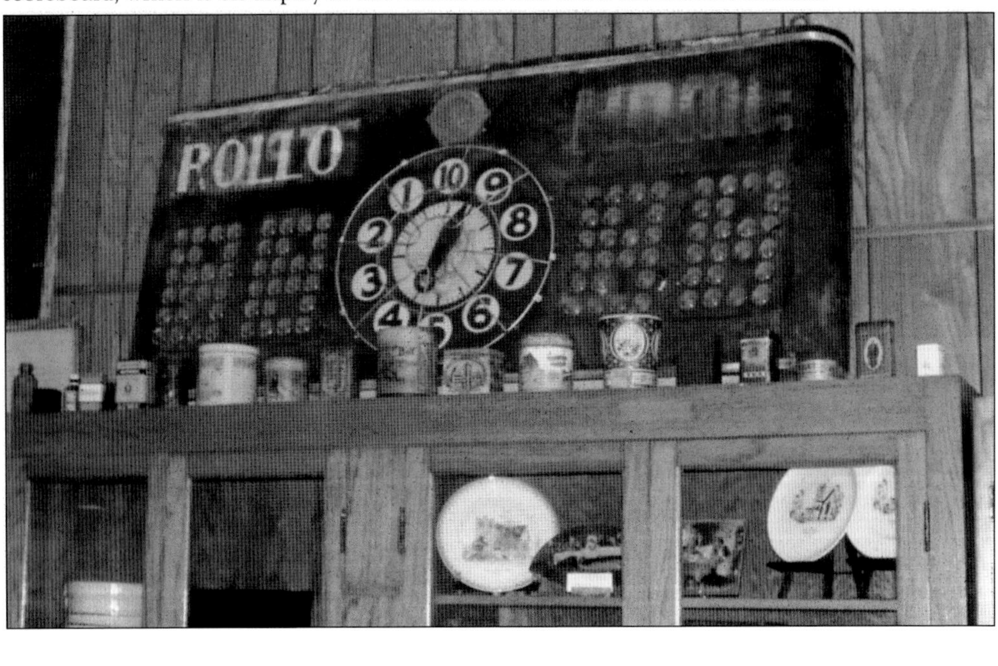

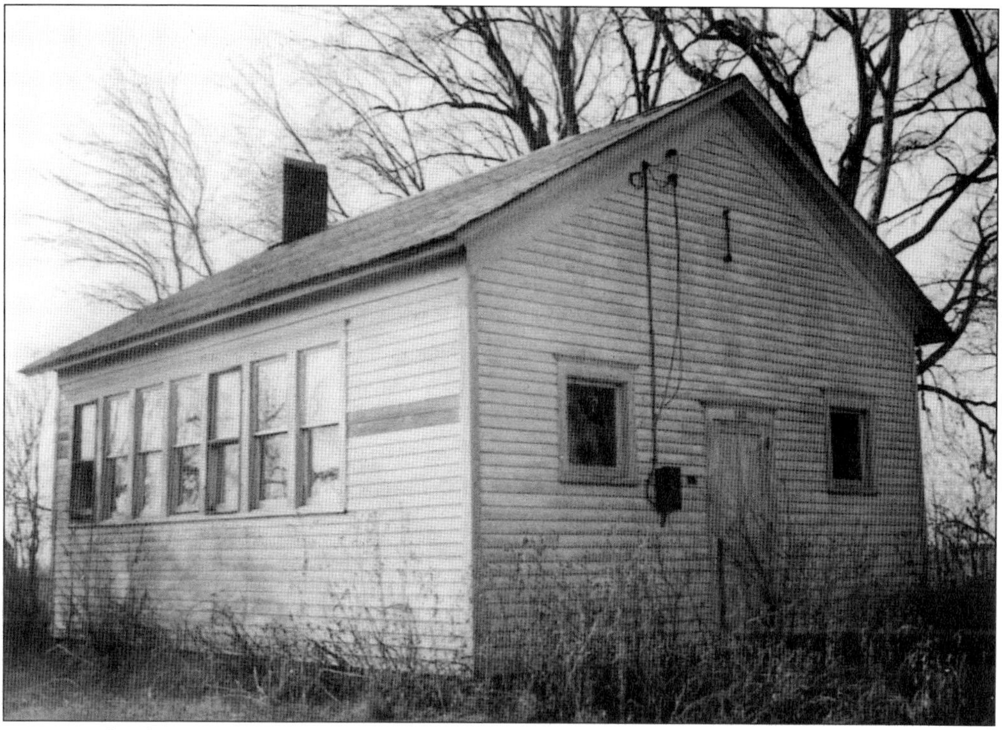

Larson School District No. 132 was organized in 1876 and 1877, and a schoolhouse was built on the south side of Sleepy Hollow Road. The Northwestern Railroad tracks were on the east side of the school, and a double row of large maple trees was on the west side. John Larson donated the land, and the first teacher was Lila Quilhot. In the early years, it was called the Foiles School. The name was changed to Larson School in the 1900s. Electricity was installed in the late 1930s, as well as an oil-burning stove. With only 11 students enrolled, the school closed in the 1950s. Land reverted back to the Larson family, and the school building was sold but never moved. Later, it burned. The picture below was taken in 1946; the teacher was Doris Larson. (Below, courtesy of Elsie Drake.)

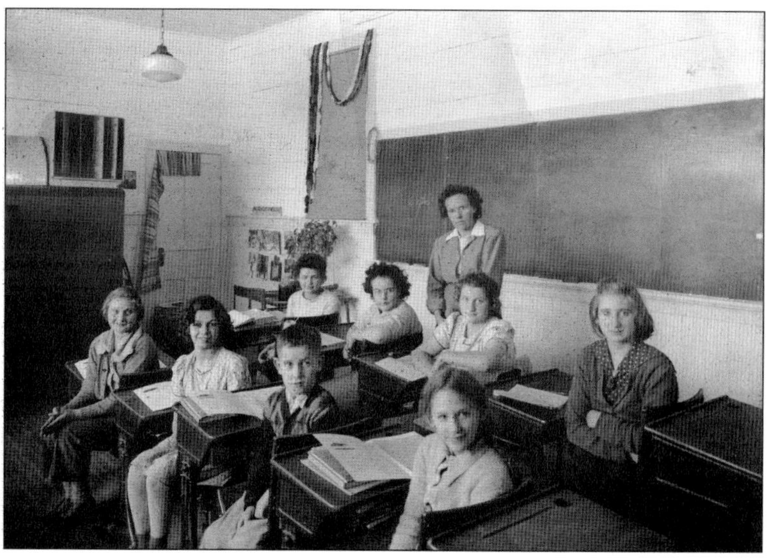

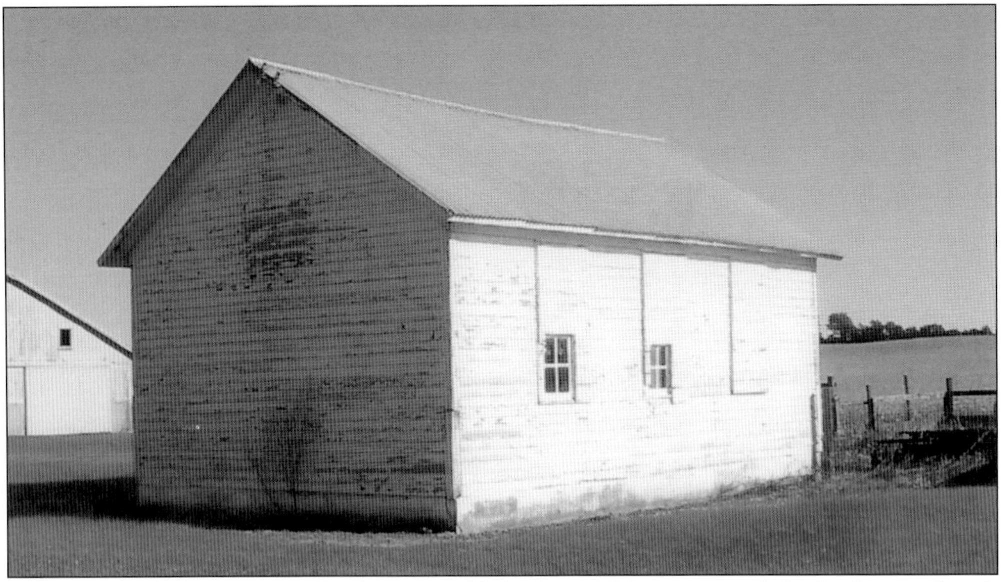

The Norwegian Parochial School was established in 1866 as a part of the Lutheran church system, and the school's objective was to teach children the catechism in the language of their fathers. The school building was located in Section 15, the southeast corner of Willow Creek Township. The 18-by-26-foot building was erected in 1881 at a cost of nearly $500, which included furniture and fixtures. Holden Risetter donated the land for the site. Children attended for two and a half months and then went into the village of Lee for a month in a public school building. The building, located on the Macklin farm southwest of Lee, is not used anymore. A member of the Risetter family, second from left in the first row, is in this 1924 picture of Lee High School. (Below, courtesy of Helen Pickwell.)

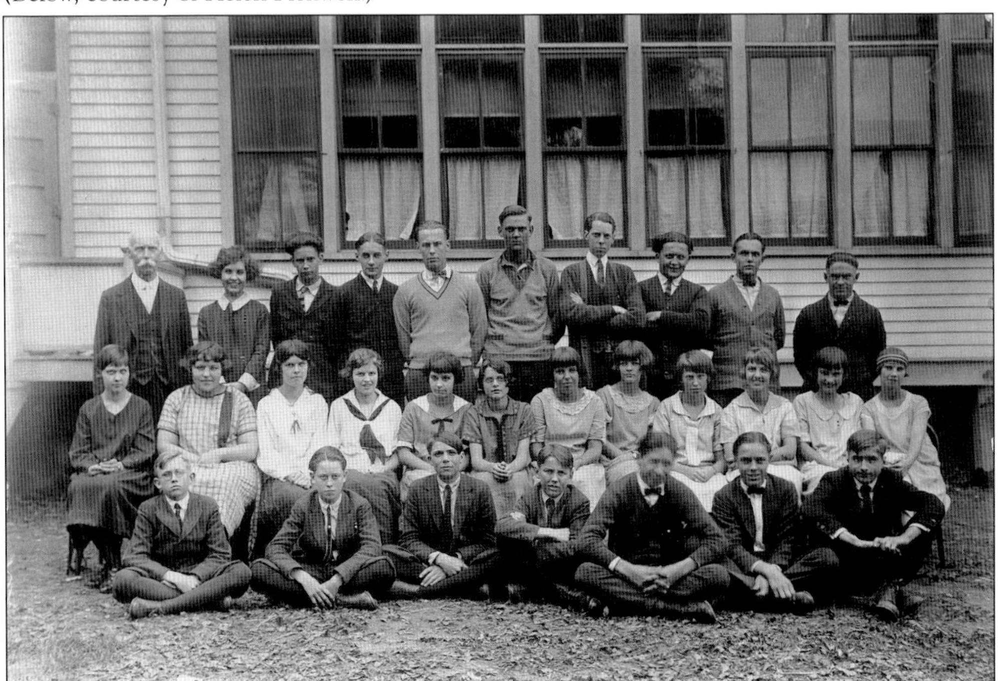

Three

Churches and Cemeteries

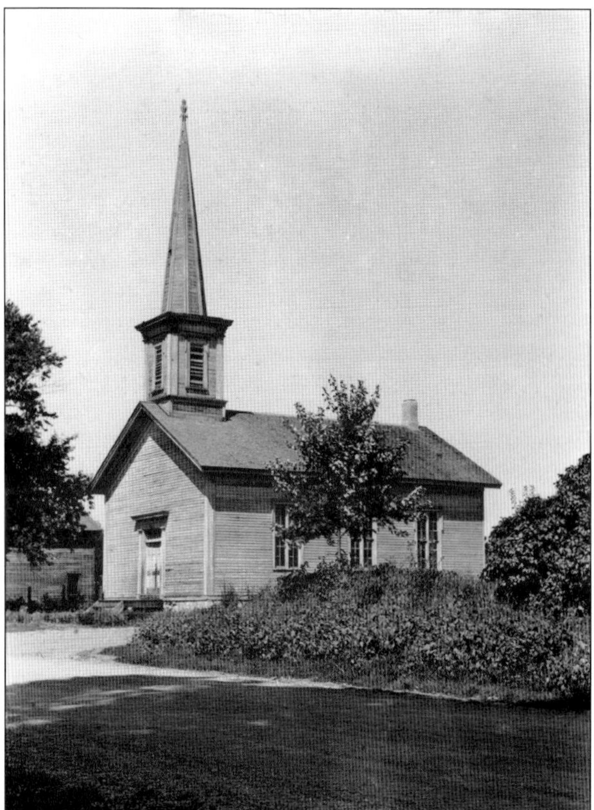

The early settlers selected Shabbona Grove for the town, as it was nestled on the southern edge of the timber. Methodist services were held at the home of Coleman Olmstead in the fall of 1841. The Methodist Church was built of logs on the northwest corner of Indian and Shabbona Grove Roads in 1864. The Baptists took over the church in 1894, and in 1946, the structure was dismantled.

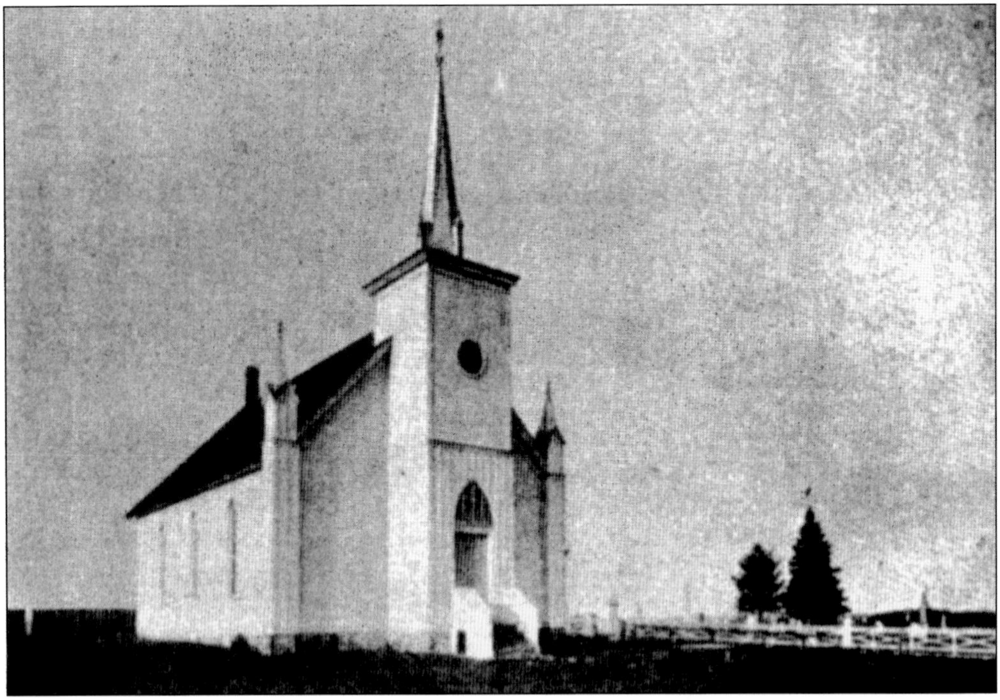

About two miles west of Shabbona, Illinois, was an English settlement, which started in the fall of 1851 when five young Englishmen took up government land. In 1857, a schoolhouse was built. Church services were held in the school until the Methodist Episcopal Church was built in 1869. Septimus Storey and his wife, Mary (Mullins), gave land in 1869 for the cemetery that was established directly east of it.

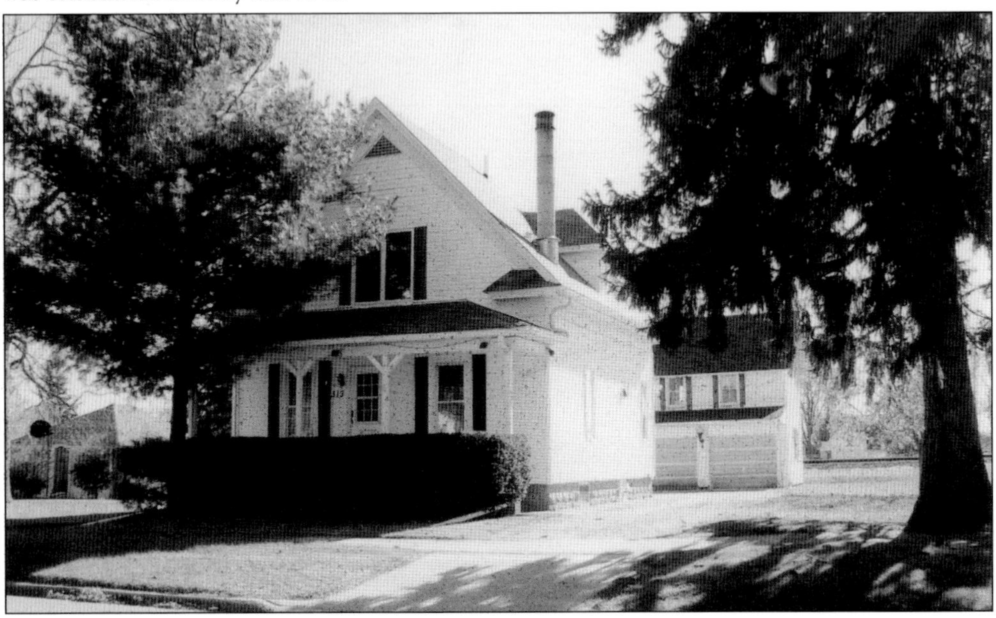

Wood from the Methodist Episcopal Church, located west of Shabbona, was hauled into Shabbona and used to build the garage at the Scott Summers residence on West Navaho Street. The front door is original, and there is a stained-glass window on the east side of the house from the church.

J.J. Meyers built the Scarboro Community Church (United Methodist) in the spring of 1875. The total cost of the church was $3,346. It was first located north of Scarboro and later moved onto the Durin property in Scarboro in 1912. When the membership declined, it was torn down. A memorial plaque for the site of the Scarboro Community Church is located on the west side of the main street in Scarboro, Illinois. The old bell from the church is also kept at this site. (Right, courtesy of Shirley Rees.)

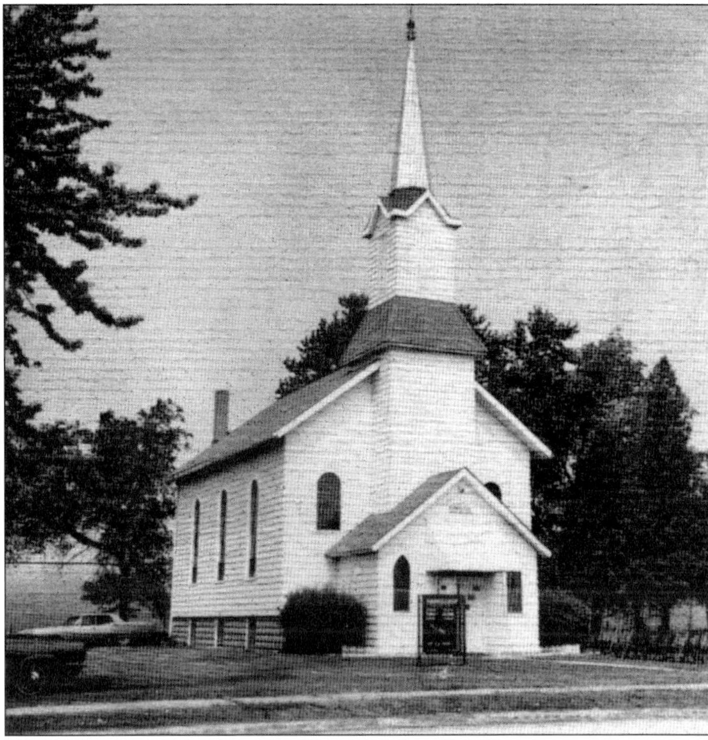

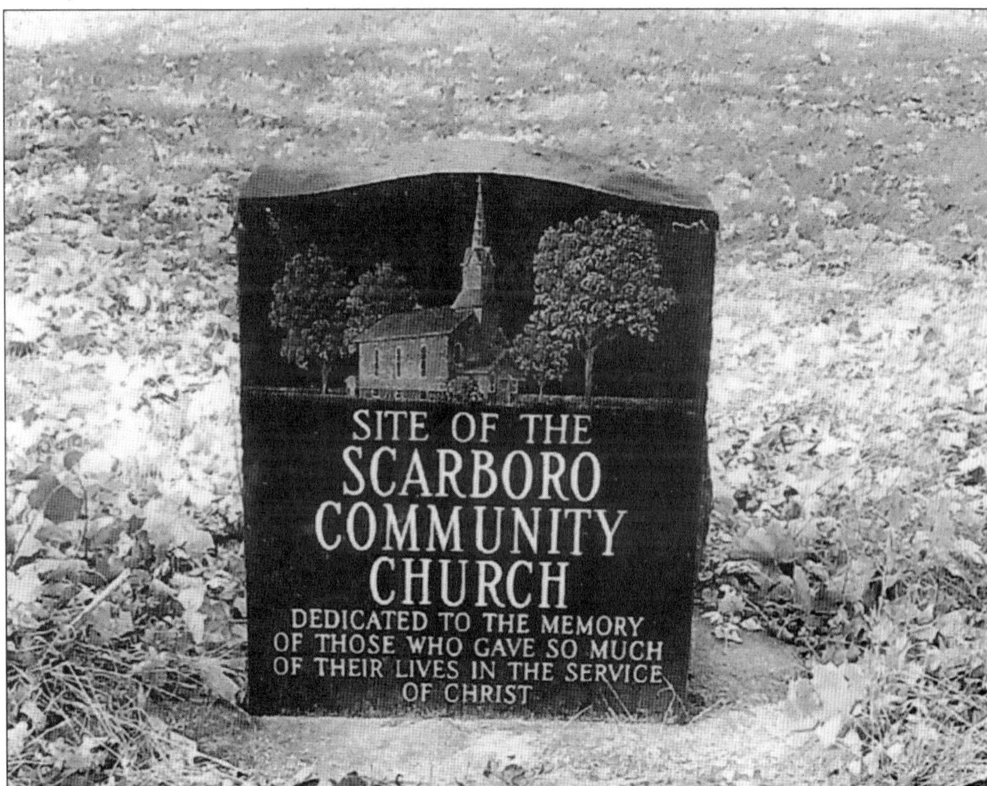

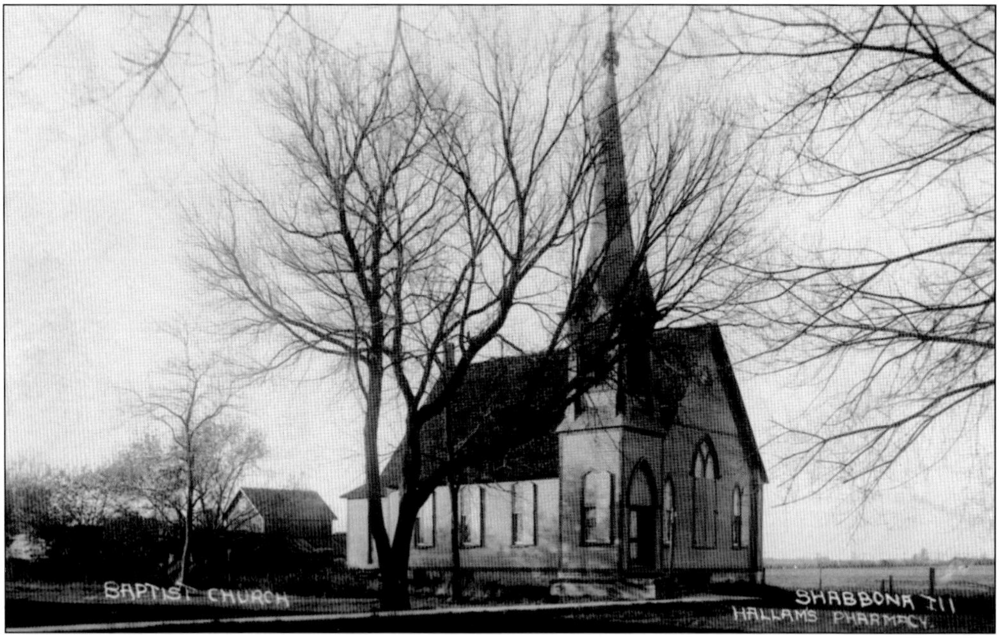

President Hinckley of the Chicago & Iowa Railroad persuaded John Rey to see what he could do toward the erection of a Baptist church in Shabbona, Illinois. After securing $1,500 for that purpose, the people in the community met in the station of the railroad and put together an organization to erect a house of worship. The First Baptist Church of Shabbona was organized on July 26, 1874, and was dedicated on November 28, 1874. A unique feature of the entryway is the frosted-glass window designed and installed by Burton Houghtby. It consists of alternating panels of blue, green, and rose; the blue portion is formed in the shape of a cross. In 2010, it became the home of the Village Bible Church. (Above, courtesy of the Hallam Family Collection.)

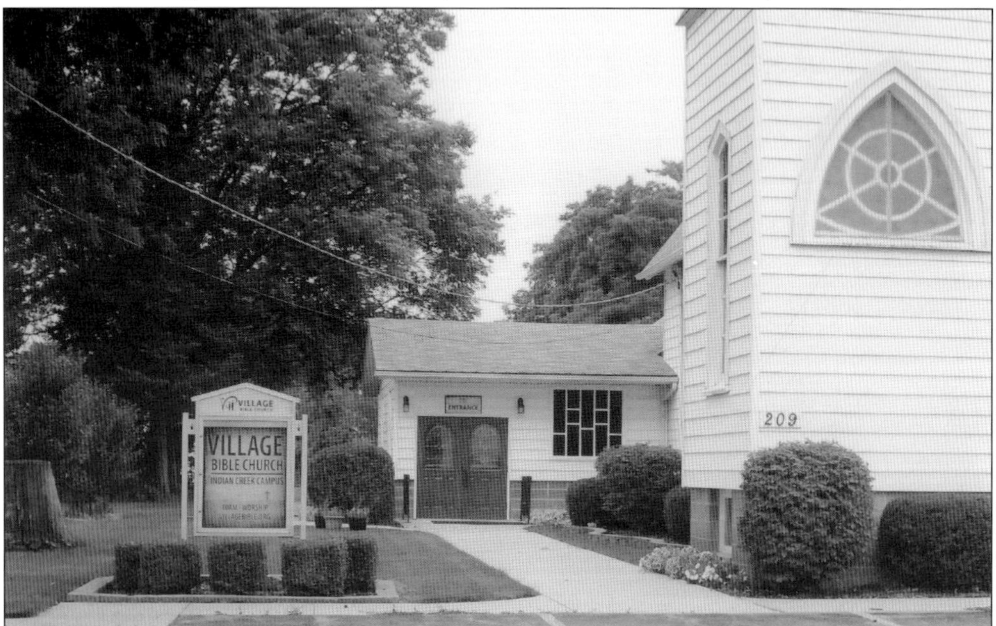

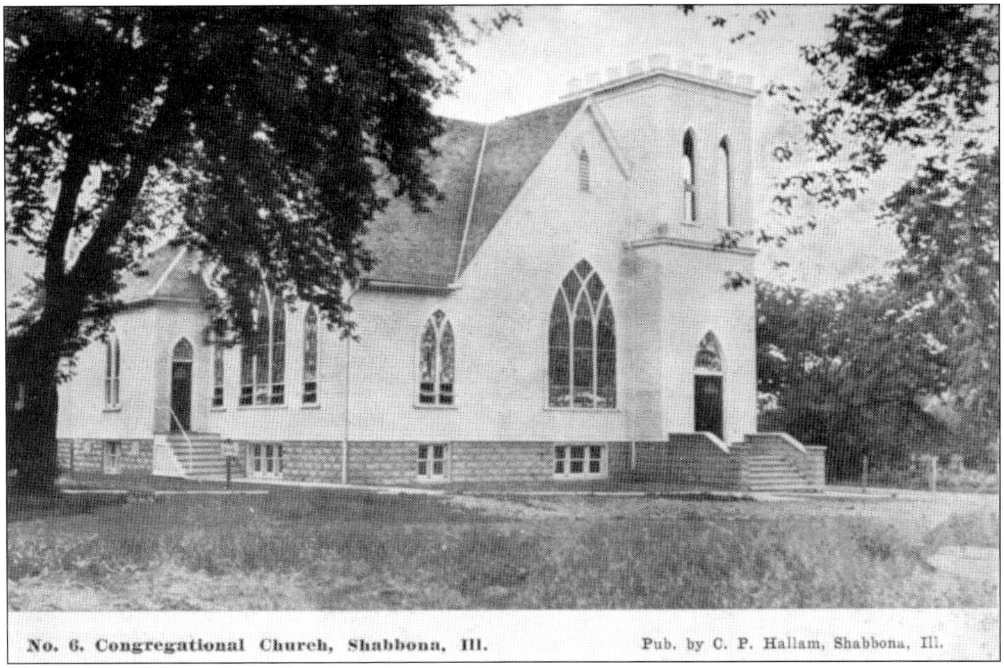

An informal Saturday evening meeting was held at Samuel Boslough's house to discuss religious needs. The Congregational Church in Shabbona, Illinois, was erected in 1879 after the congregation had used the Baptist building for services since 1876. In 1914, services were discontinued at the Old English Methodist Church west of Shabbona, and a big part of that congregation united with the Congregational Church. A new constitution was approved in July 1961, which officially changed the name of the church to the First United Church of Christ in Shabbona. A new educational unit was completed and dedicated on October 13, 1963. In 1967, a new parsonage was built, and a new church steeple was erected in May 1975. (Both, courtesy of the Hallam Family Collection.)

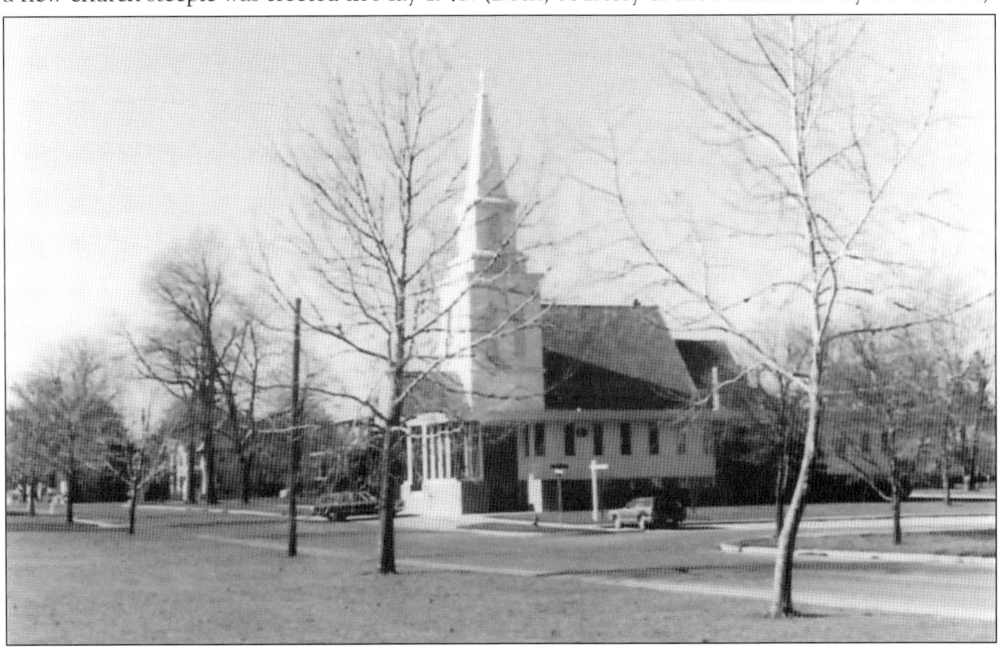

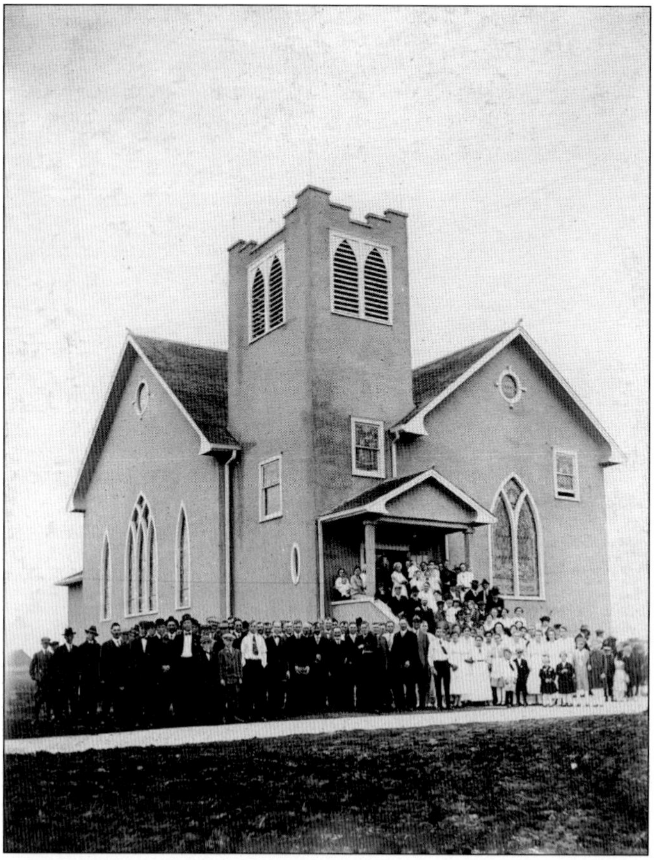

The Calvary Lutheran Church on Perry Road in Lee, Illinois, was started in the community by a group of Norwegian settlers who came to the Lee area around 1860. They met in homes for religious services. In 1870, this group of settlers organized the Norwegian Lutheran Congregation, and Reverend Torgerson was the first pastor called. In July 1871, plans were begun to build a church, and Elias Espe offered the congregation an acre of land one mile west of the cemetery. The building was completed in 1874 and destroyed by lightening in 1914. A new building, below, was erected in 1917, with an educational unit added in 1963 and a bell tower in 1970. The parsonage was built in 1948 on land donated by Melvin Jacobs. The Union Cemetery is located west of the church. (Left, courtesy of Bill Espe.)

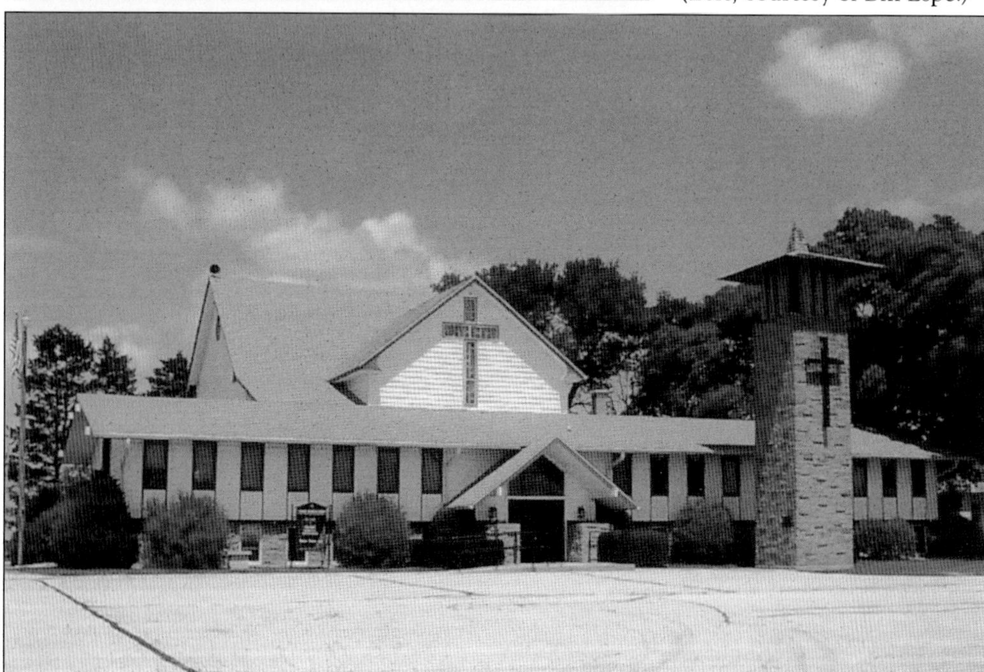

The First Lutheran Church in Lee, Illinois, started in October 1858 when six families living in Bradford Township, Lee County, met to organize a congregation of the Lutheran faith. All of the families came from Kinservig's Parish, Hardanger, Norway. The property for the first cemetery and church, located two and a half miles southwest of Lee, was acquired from Haldor Eden in 1861. It was completed and dedicated in 1868; a new building was erected in 1896 at the same site. It was moved to its present site in Lee in the summer of 1924, and the cornerstone was laid on October 4, 1925. A renovated building (1948) was purchased across the street from the church for the parsonage. The First Lutheran Cemetery is located on Lee Road.

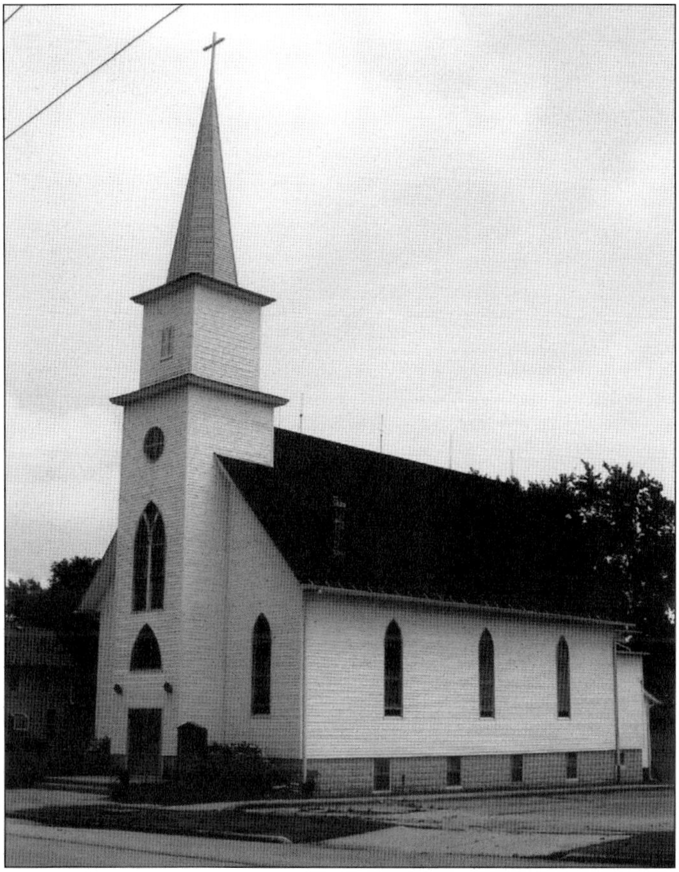

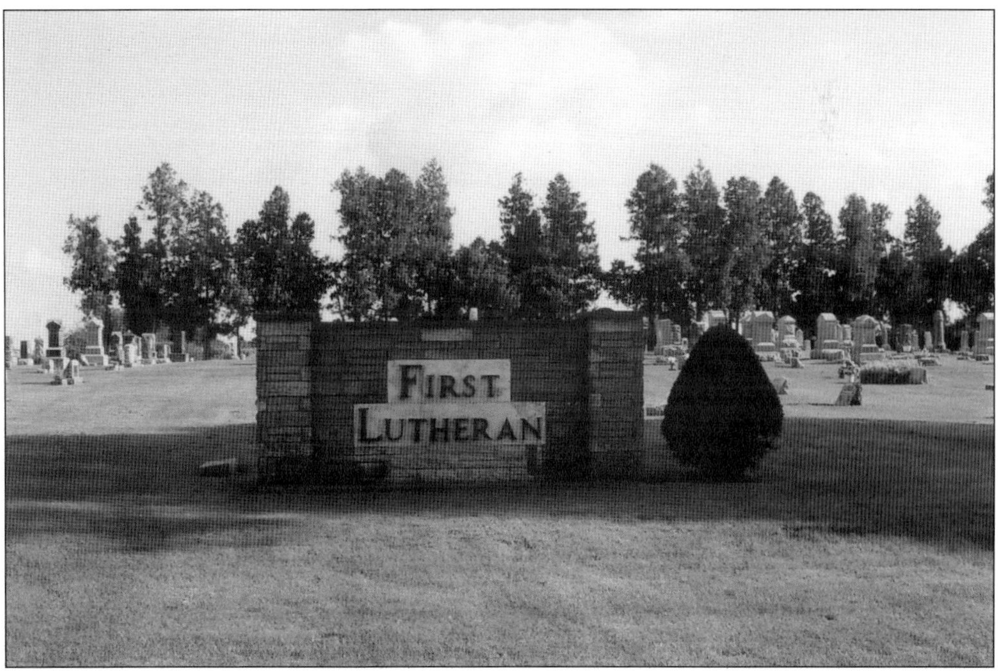

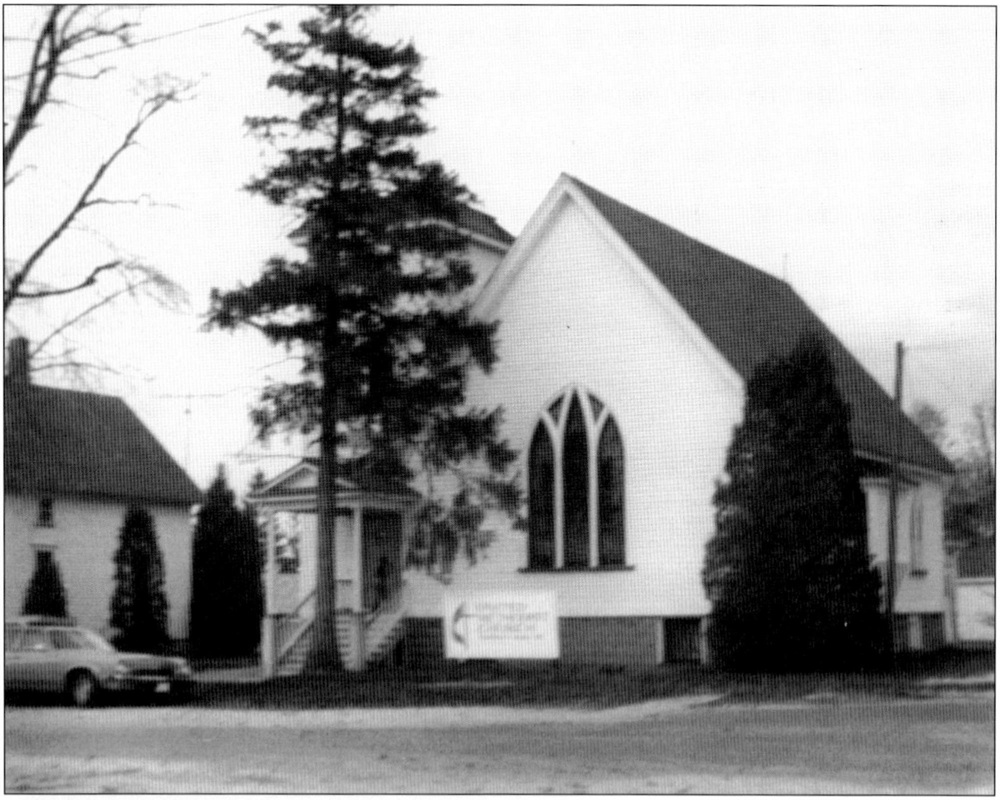

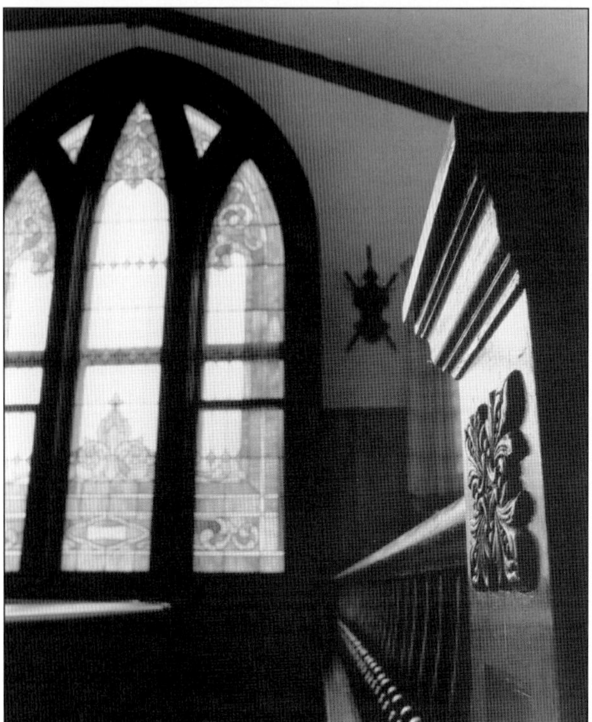

The Lee United Methodist Church was founded in 1841 when Coleman Olmstead came to Shabbona Grove and built a home and barn, which were used to hold circuit rider meetings. Several English families established a settlement west of Shabbona, and a church was built in 1869. In 1876, Francis Hinckley of the Chicago & Iowa Railroad sold the Methodists land in Lee for a church for $5. A Mr. Blee built the church for $95 on the east side of Lee. In 1895, the Hendersons sold their home next to the church to the Methodists for the sum of $100 for a parsonage. The books of the Old English Church were turned over to the Lee Methodist Episcopal Church in 1902. Several improvements were made to the building until it eventually closed and became a private home. (Both, courtesy of Alice Lintereur.)

The St. James Catholic Church was built four years after Lee was incorporated in 1874, with pioneers John Kennedy, James Kirby, M.P. Harris, and Bernard Malloy developing the plans for the church. On October 13, 1875, John Kennedy donated an acre of land, and the quartet raised $3,000 from area Catholics. The church was built under the supervision of Fr. Edward Froelich, the pastor of Saint Patrick in Rochelle, Illinois, and remained a mission parish for almost 30 years. In 1949, the shelter sheds for horses burned and threatened the church. However, the church remained and has since been renovated. The Social and Education Center was dedicated in 1973. The St. James Rectory sits beside the church, and the St. James Cemetery is located on County Line Road. The picture below was given to the museum by Margaret Duffy and shows the inside of the Catholic church.

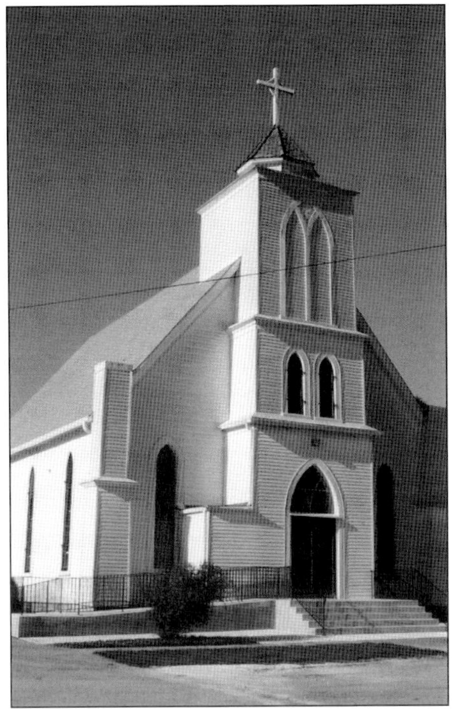

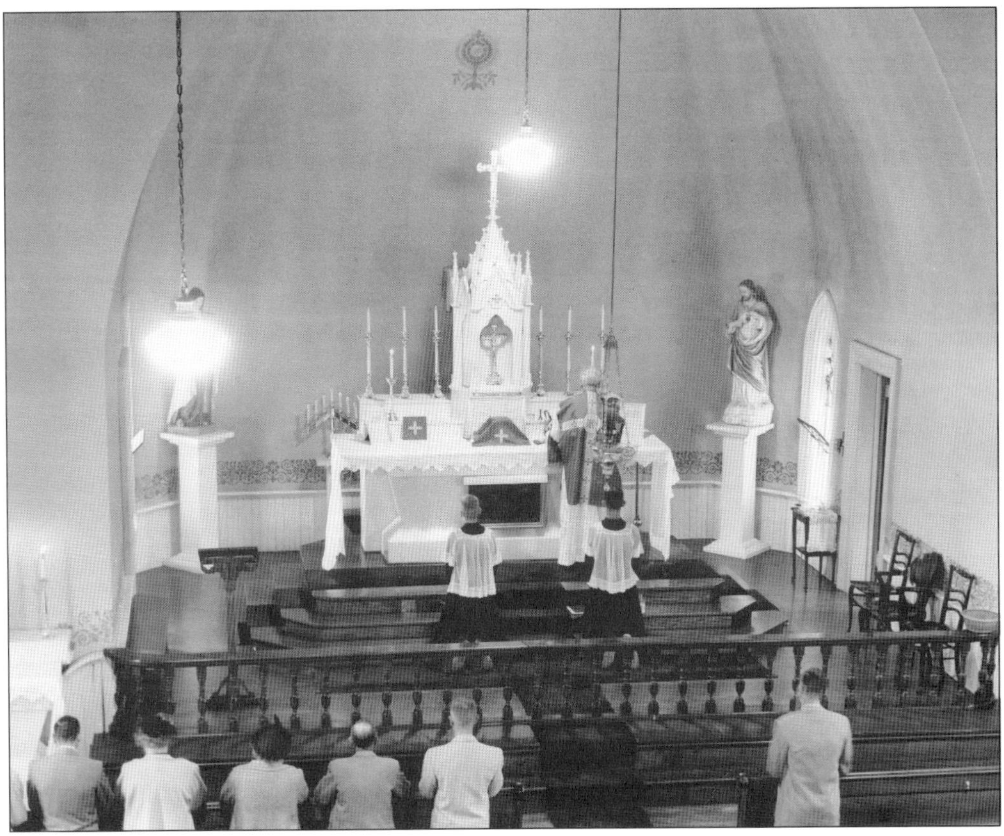

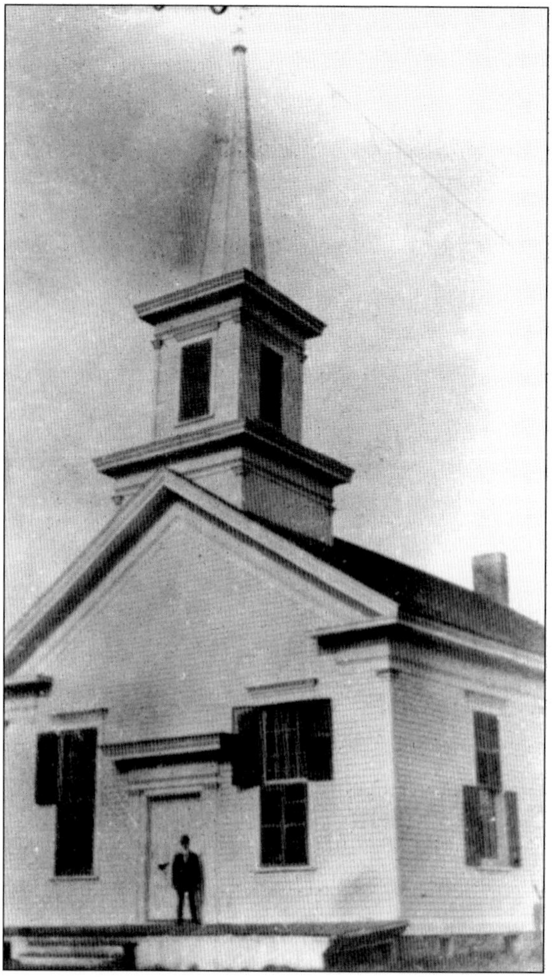

The second organized religious group in Paw Paw Township is what is now known as the United Church of Christ (Congregational Church) in Rollo. It was originally called the First Independent Union Church of Paw Paw Township. The first church building east of Rollo on Hyde Road was erected and dedicated in 1867. In 1876, the number of members was 50. Thinking to further the religious interest of the church and community, the members moved the church building and parsonage to Rollo in the fall of 1886. Because of damage from a windstorm, a new church, shown below, was built at a cost of approximately $14,000 and dedicated on June 21, 1914. The parsonage was later destroyed by fire in the summer of 1894. On August 19, 1914, a new brick parsonage, which cost $5,000, was dedicated.

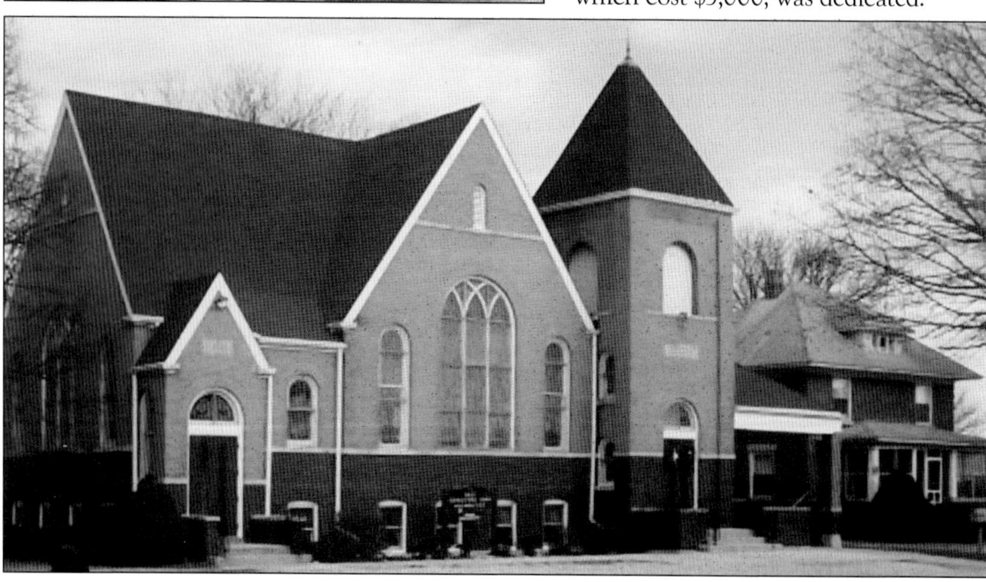

Four

TRANSPORTATION AND COMMUNICATION

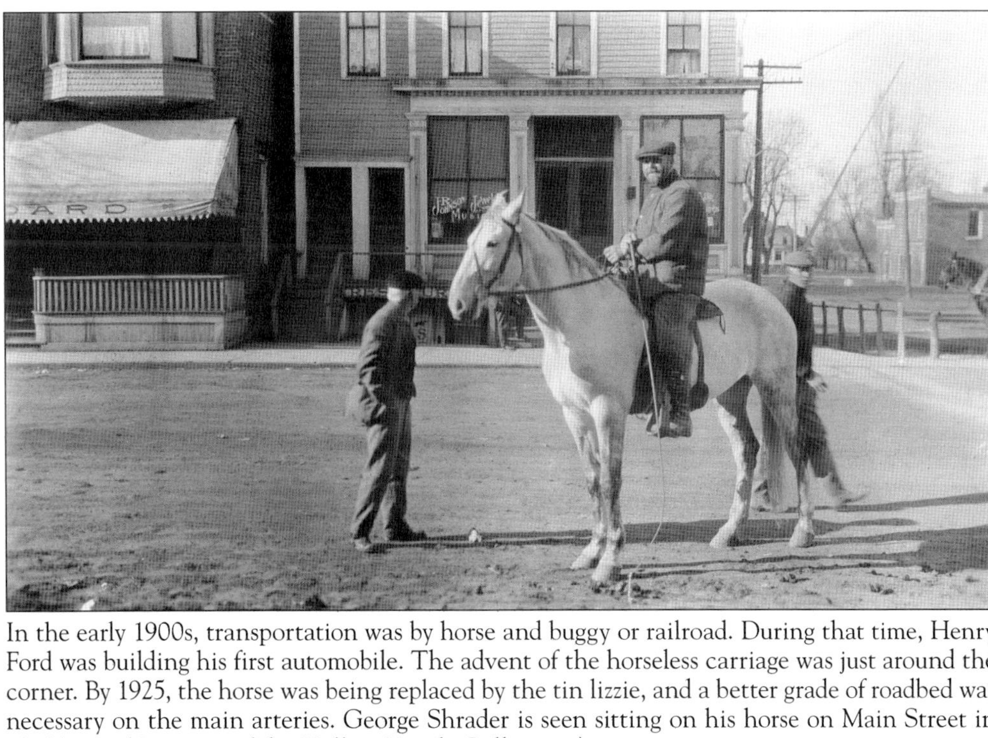

In the early 1900s, transportation was by horse and buggy or railroad. During that time, Henry Ford was building his first automobile. The advent of the horseless carriage was just around the corner. By 1925, the horse was being replaced by the tin lizzie, and a better grade of roadbed was necessary on the main arteries. George Shrader is seen sitting on his horse on Main Street in Shabbona. (Courtesy of the Hallam Family Collection.)

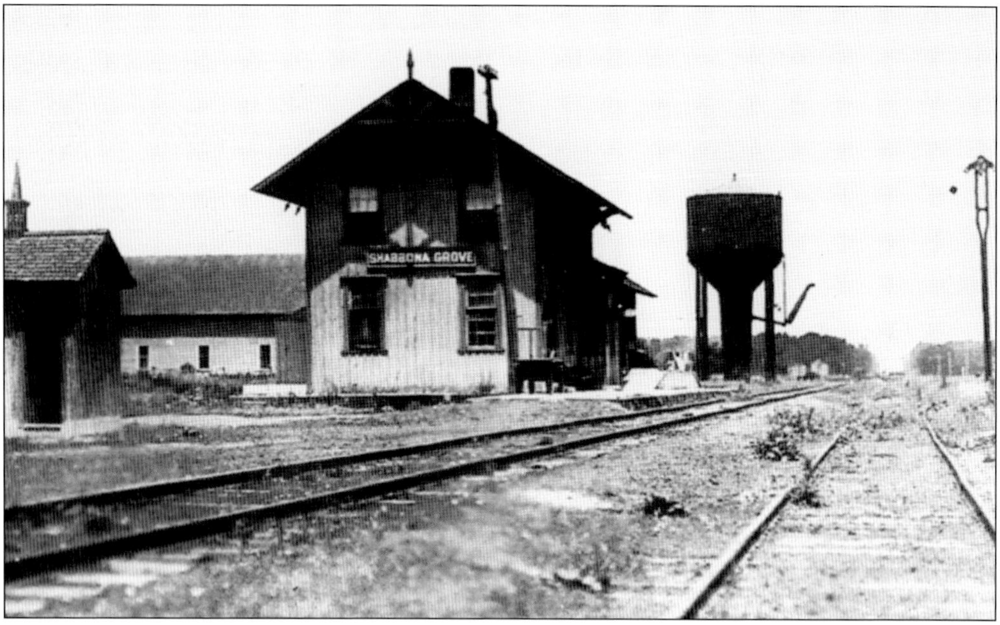

In the early 1920s, the railroad depot was the most southern building in Shabbona Grove. In the best days, Shabbona Grove had three general stores, a tin shop, two hotels, two churches, a Masonic lodge, and a boot and shoe shop. The Olmstead brothers owned and operated a sawmill. Reuben Challand ran a blacksmith shop. The Chicago & Northwestern Railroad received permission to close its station at Shabbona Grove on the Spring Valley line in 1942. Train mail service had been discontinued in 1929. The Chicago & Northwestern Railroad ran a train from DeKalb to Spring Valley. Clarence Todd took this locomotive picture (below) from Shabbona Grove Road on the Spring Valley Line.

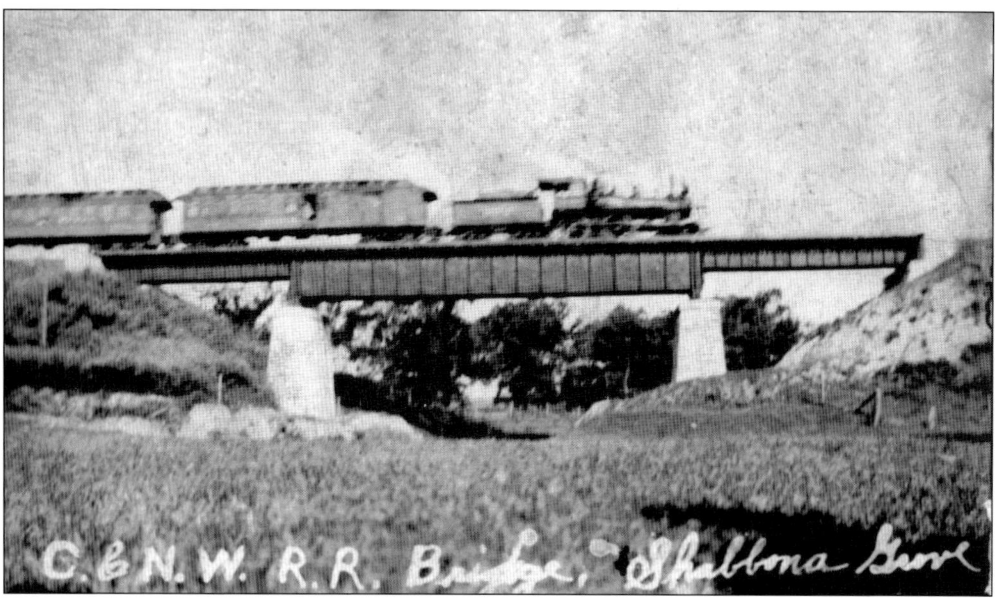

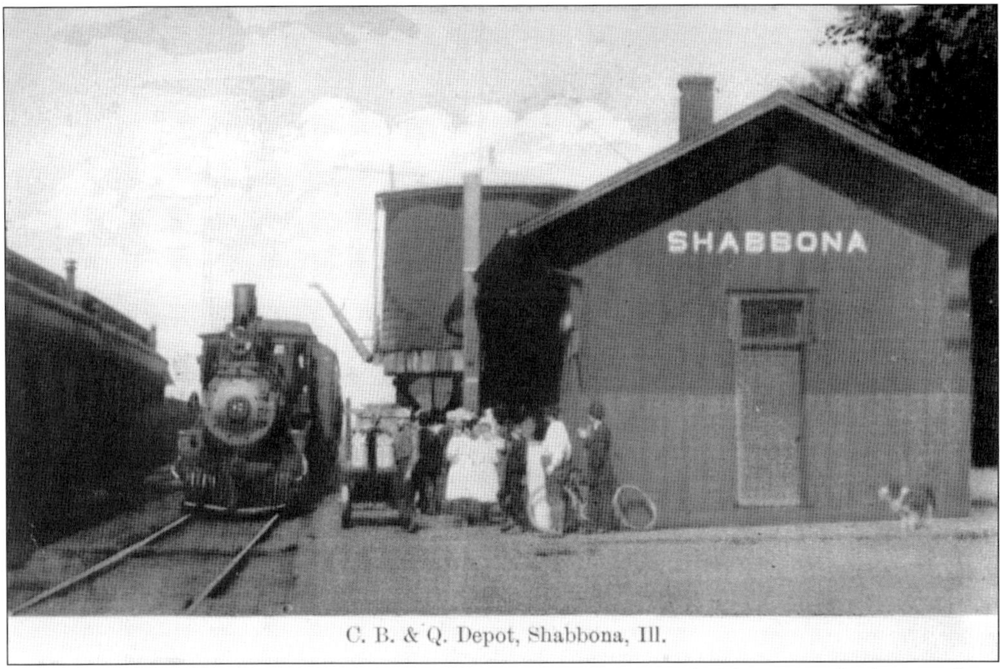

In 1873, Shabbona Depot was moved to its present location. Shabbona reached its peak in railroading in the early 1900s. At this time, they boasted a two-engine roundhouse and three switch tracks. Shabbona was also the source of a milk train. It started in Shabbona and hauled milk into Chicago, making several stops along the way. The train transported some passengers. Note the milk cans ready to go. (Courtesy of the Hallam Family Collection.)

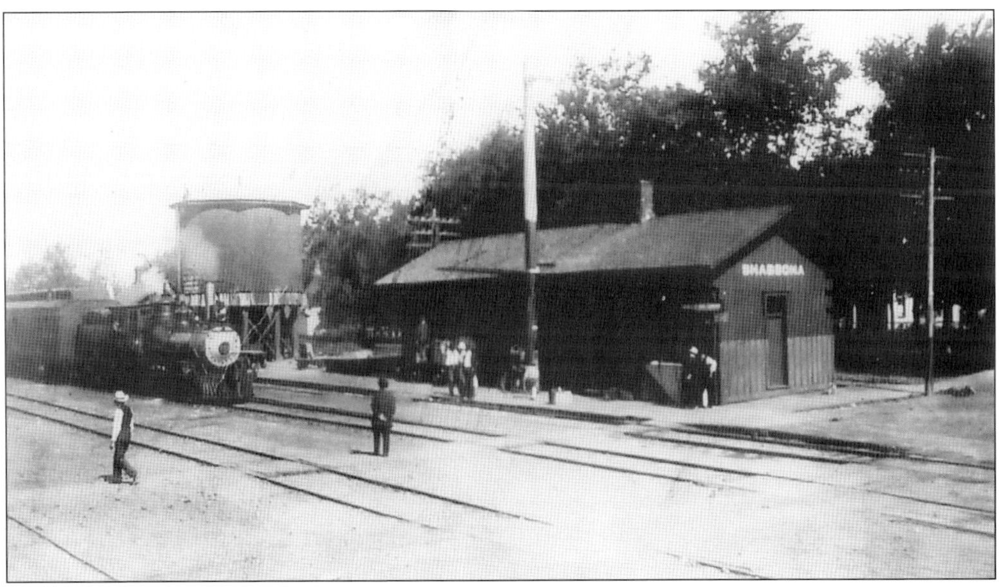

Soon after building the Chicago & Iowa Railroad through Shabbona, a branch that went from what was known as the "Y" to Paw Paw was put through. The main purpose of this road was to haul coal up from the southern part of Illinois. The early photograph here is a steam engine filling with water at the Shabbona Depot. (Courtesy of the Hallam Family Collection.)

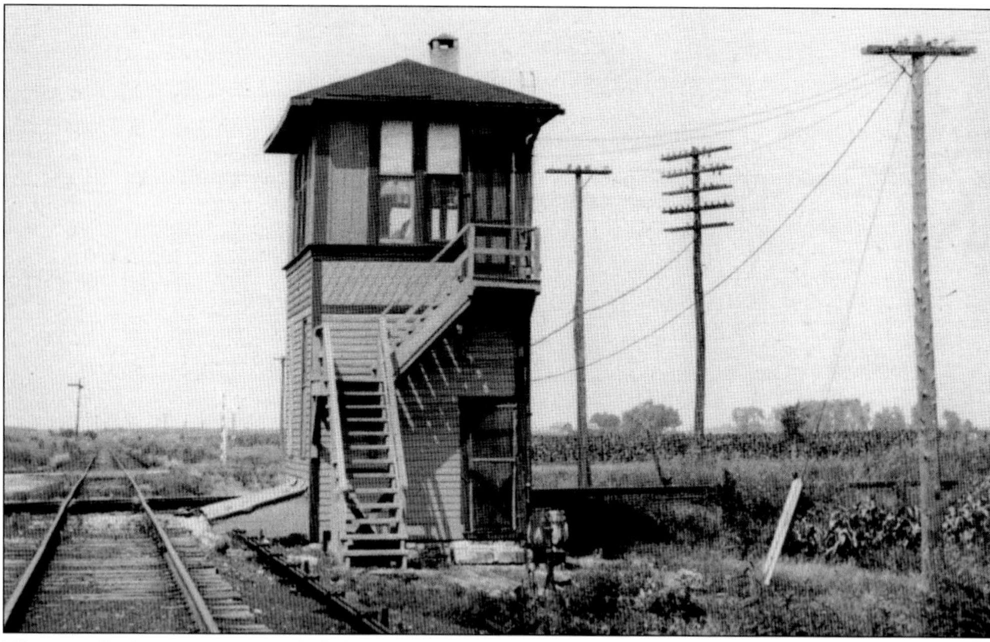

The interlocking tower guarded the intersection of the Chicago, Burlington & Quincy (CB&Q) Aurora-Savanna main line and the Chicago & North Western's Sycamore Spring Valley and Seatonville, Illinois, branch. This August 25, 1937, photograph looks northward along the Chicago & Northwestern tracks; the Chicago, Burlington & Quincy main line runs from left and right across the photograph, and today's US 30 is seen just beyond. This picture of the tower, located between Waterman and Shabbona, was given to the museum by Helen Pickwell.

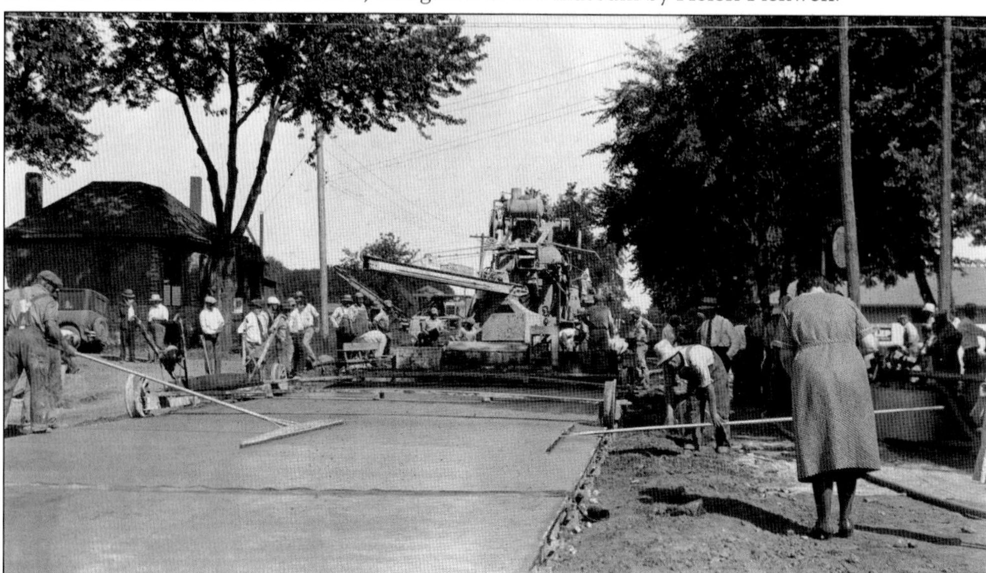

According to the *Shabbona Express* of July 17, 1930, work was actually started on Route 71 from Waterman to Shabbona. Upon completion, the cement road ran east from Shabbona to Aurora and on to Chicago. A celebration was held on September 28 and 29, 1932. The viaduct crossing the CB&Q Railroad was completed in late 1936. In 1938, Route 71 became Route 30. One can still drive from coast to coast on this historic road.

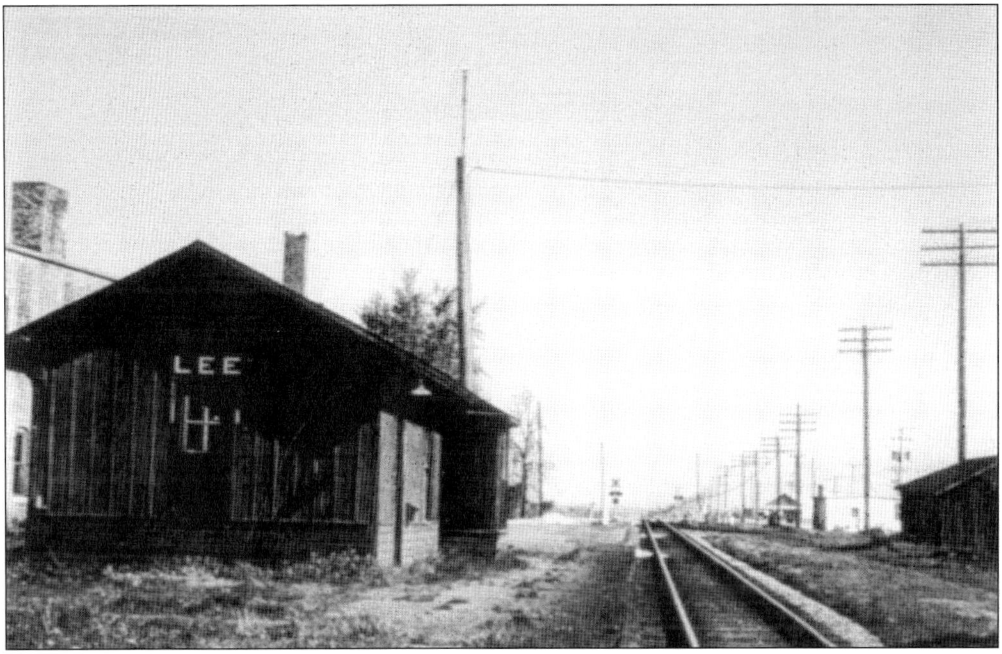

In 1899, the Chicago & Iowa Railroad was conveyed to the Chicago, Burlington & Quincy Railroad Company by deed. Joe Rambo came to Lee and served the railroad for 52 years. He retired at the age of 80 and was the oldest man in Burlington service at his retirement. Anna Wilson Arndt was night operator at the depot and one of the first lady dispatchers in the country.

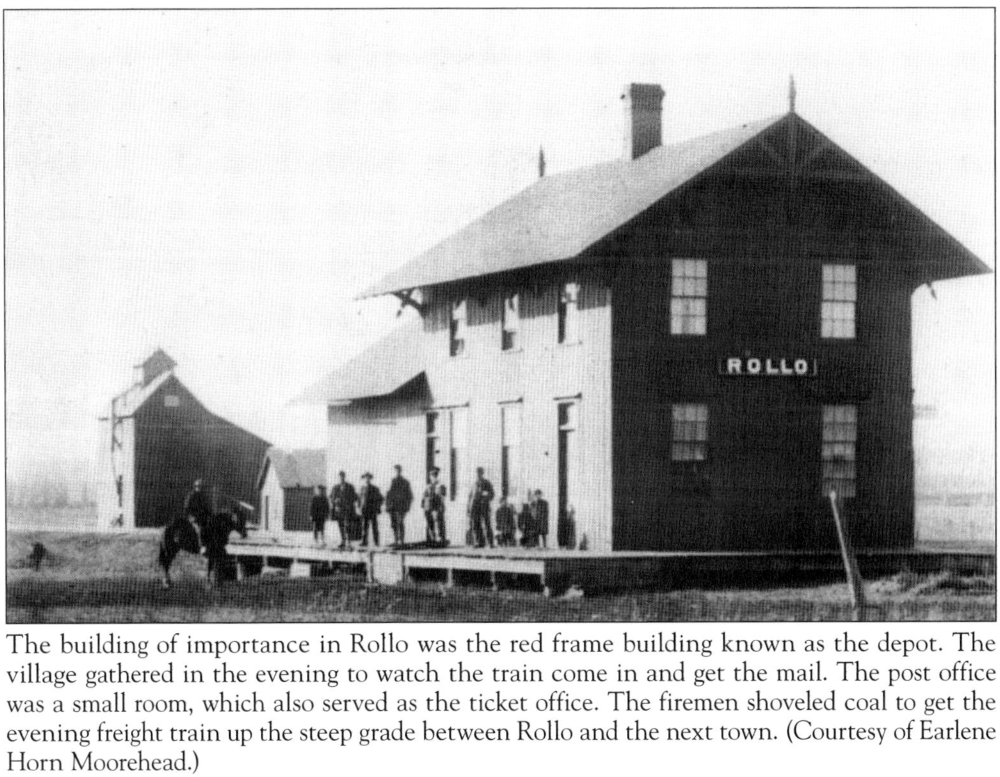

The building of importance in Rollo was the red frame building known as the depot. The village gathered in the evening to watch the train come in and get the mail. The post office was a small room, which also served as the ticket office. The firemen shoveled coal to get the evening freight train up the steep grade between Rollo and the next town. (Courtesy of Earlene Horn Moorehead.)

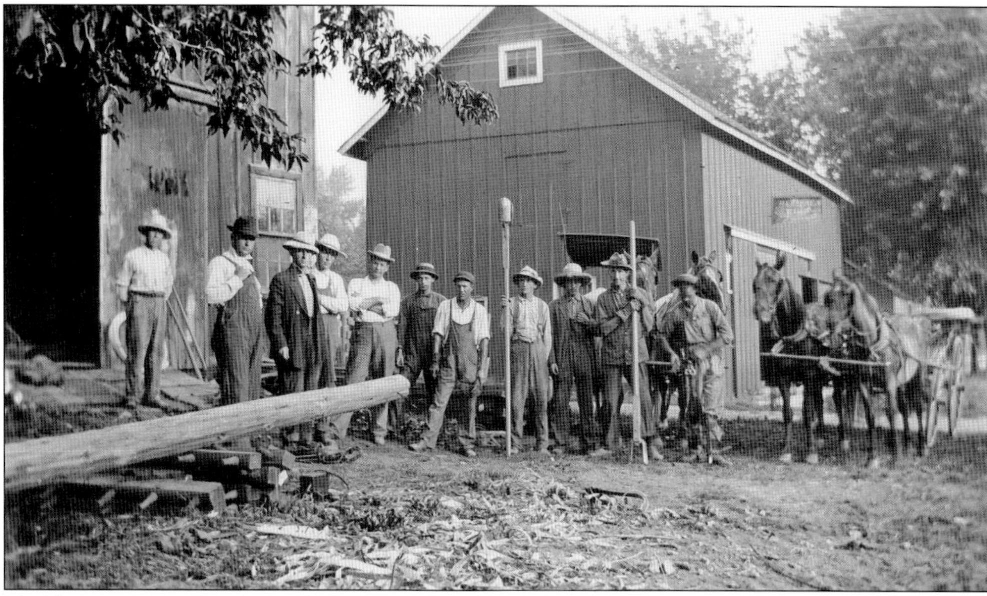

Shabbona received its first telephone service in 1895. In 1912, residents of Shabbona and other neighboring communities banded together to form their own telephone company. In 1919, the company was sold to DeKalb Ogle Telephone Company. Shabbona's dial service was established in 1937. The photograph shows men putting in the first telephone poles behind Sarah Carlson's house in Shabbona. (Courtesy of Ava Challand.)

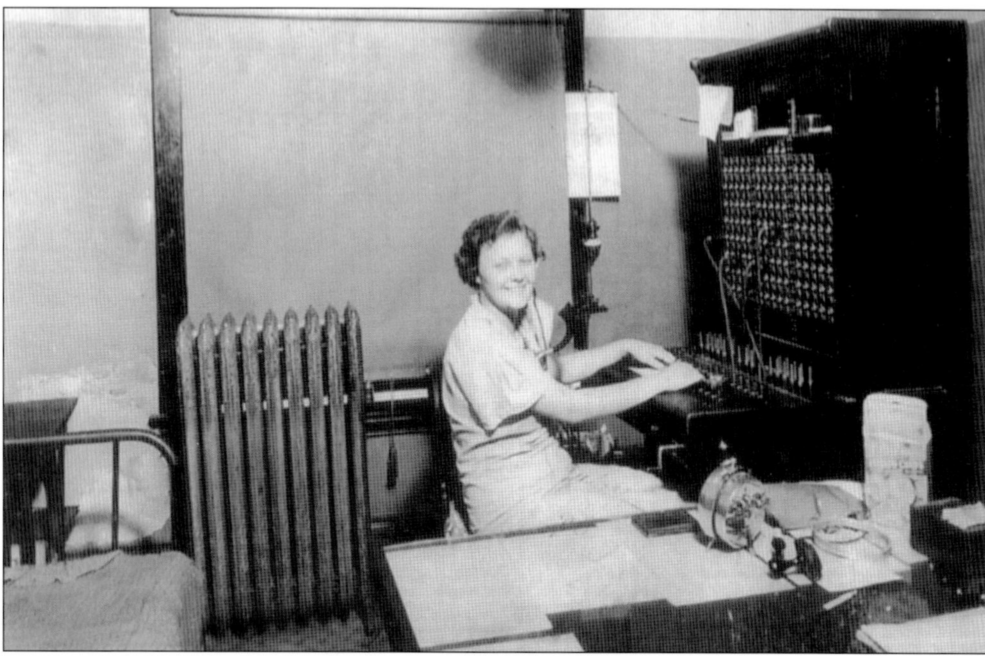

Shabbona Central Telephone Office was located on the top floor of the bank building next to Hallam's Pharmacy on Main Street in Shabbona. The operators actually made the connections on the telephone by way of the switchboards. It is amazing to consider that in the past every telephone call had to be routed manually by operators like the one pictured above. The operator is Darlene Osland, shown around 1935.

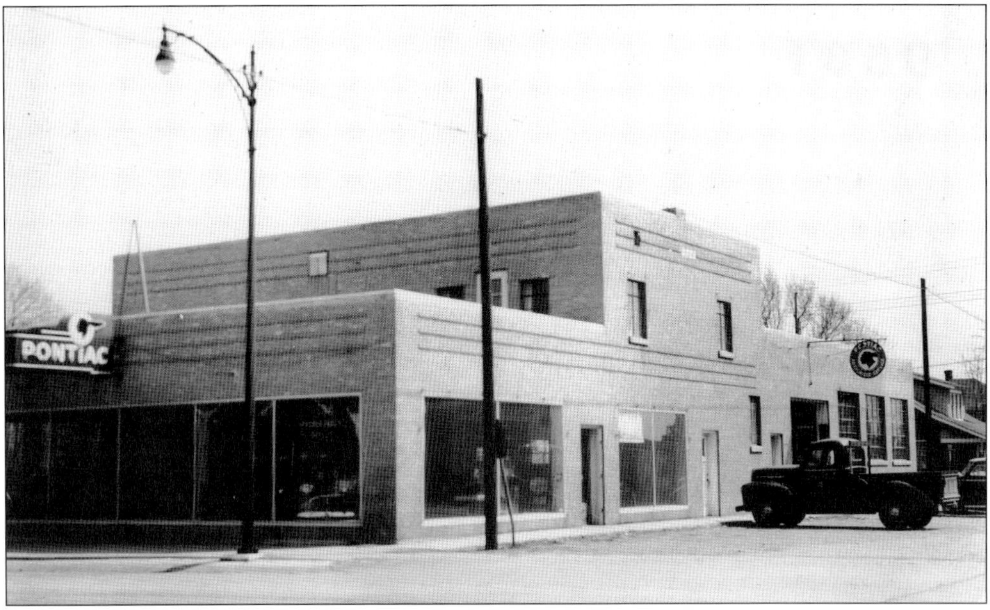

In 1918, Stanley Firkins was a sergeant in Company K, 40th Infantry, and 14th Division in World War I. He was born in 1899 in Rollo, Illinois, and came to Shabbona in 1936, purchasing Firkins Motors (Ford and Lincoln cars). Stanley Firkins Pontiac Garage supplied cars for driver training classes and buses for Shabbona High School in 1938 and 1939. The picture below is of a 1938 Chevy used for training drivers at Shabbona High School.

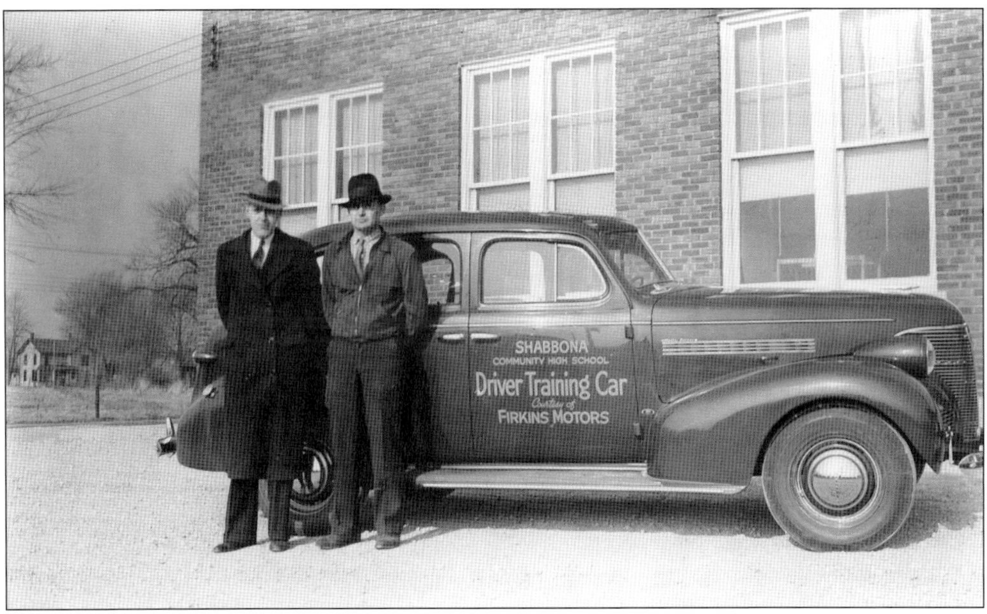

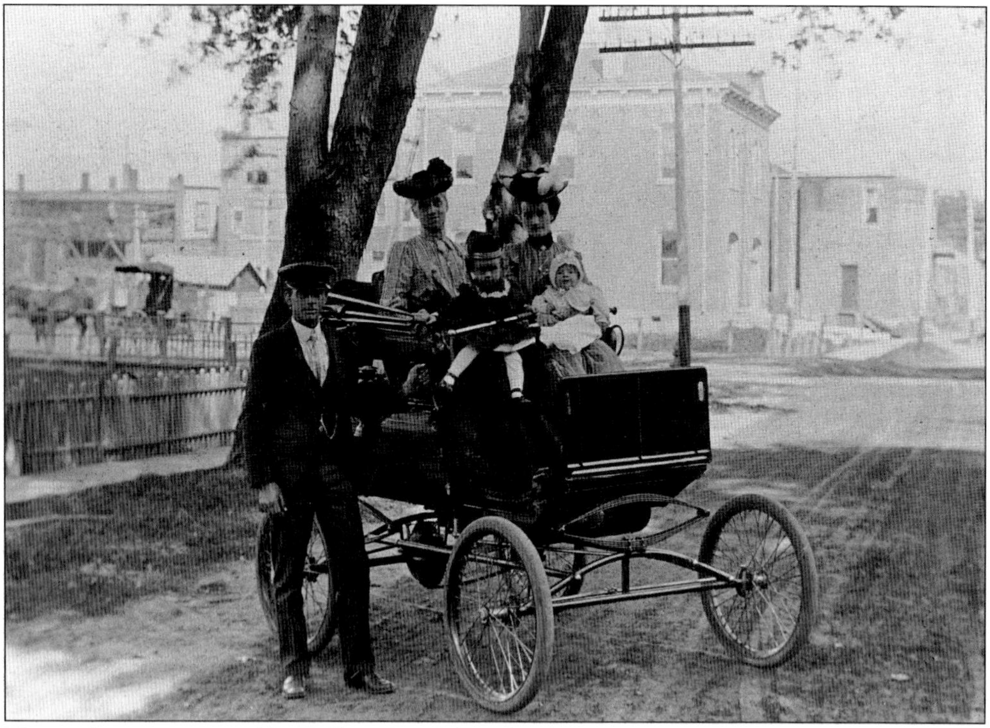

This early car was owned by William Bend (standing), grandfather of David Bend of Shabbona. On the right, Lura is holding her baby Harold. Great-grandmother Arolia is holding Evelyn, the oldest daughter. The family lived on Bend Road, south of Shabbona. Note the stick used for steering, which is found across Evelyn's lap, and the grandfather is holding on to the brake. This photograph was taken in 1903. (Courtesy of Carol Bend.)

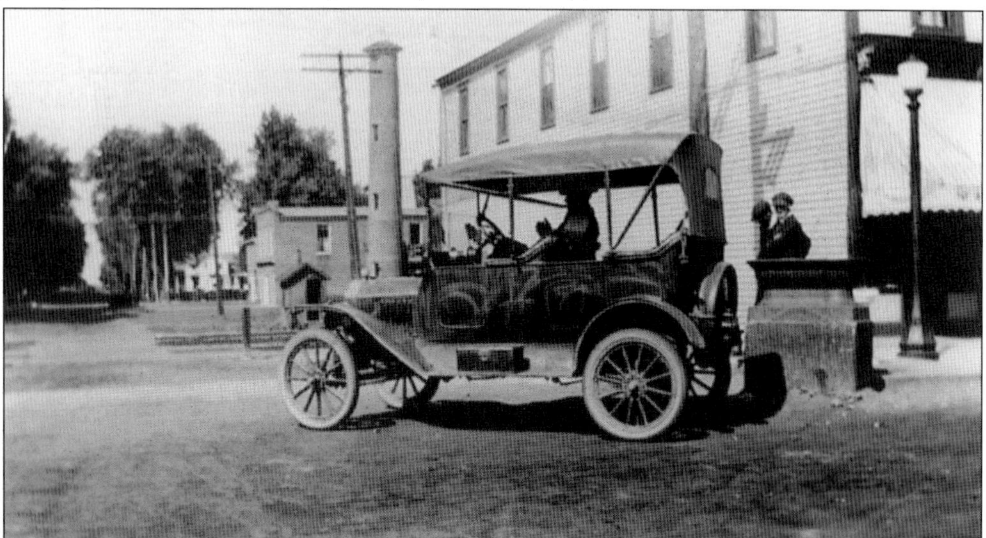

Otto Schmidt is in his car at the corner of Shabbona's State and Madison Streets (Illini and Comanche Streets). Note to the right of the car are the cast-iron trough for thirsty horses and a water hydrant, which remained until the street was paved in 1932. Schmidt was very active in the village, serving on the village board and as mayor for many years.

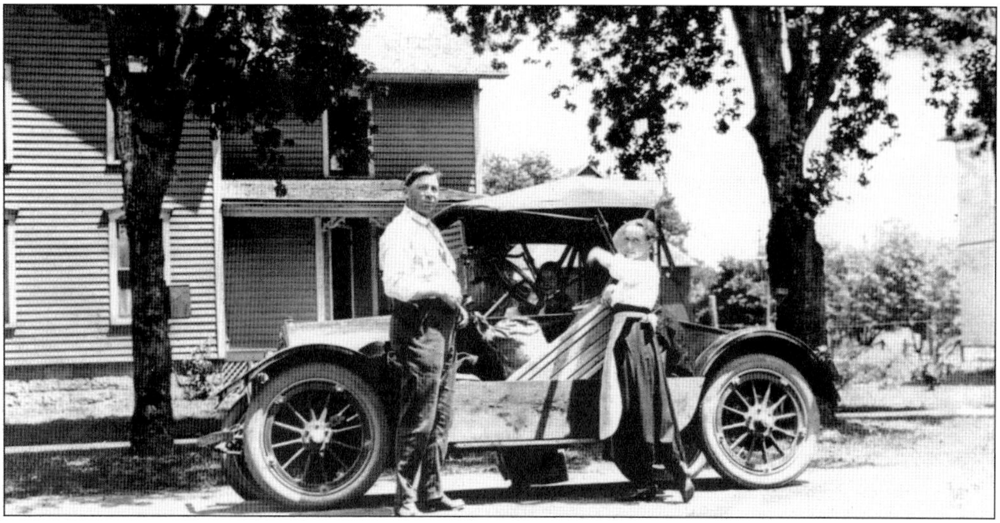

By 1900, Shabbona was 30 years old. Transportation was by horse and buggy or railroad. George Whitford, a barber, owned the first automobile in Shabbona. He boarded a train for Chicago, where he bought a 1904 Cadillac. He drove it to Aurora and stayed overnight in a hotel. Members of the Hallam family in Shabbona owned the early roadster above. The 1950 Mercury below is owned by LaVern and Marjorie Olson of Shabbona. In 1961, they bought their home at 102 East Navaho Street (in the background) from Robert Iftner. The Wormley sisters, who had moved their house from Shabbona Road because they wanted to live next to the church, were the owners of the property. Olson was a farmer and had a corn-shelling business in the area for many years.

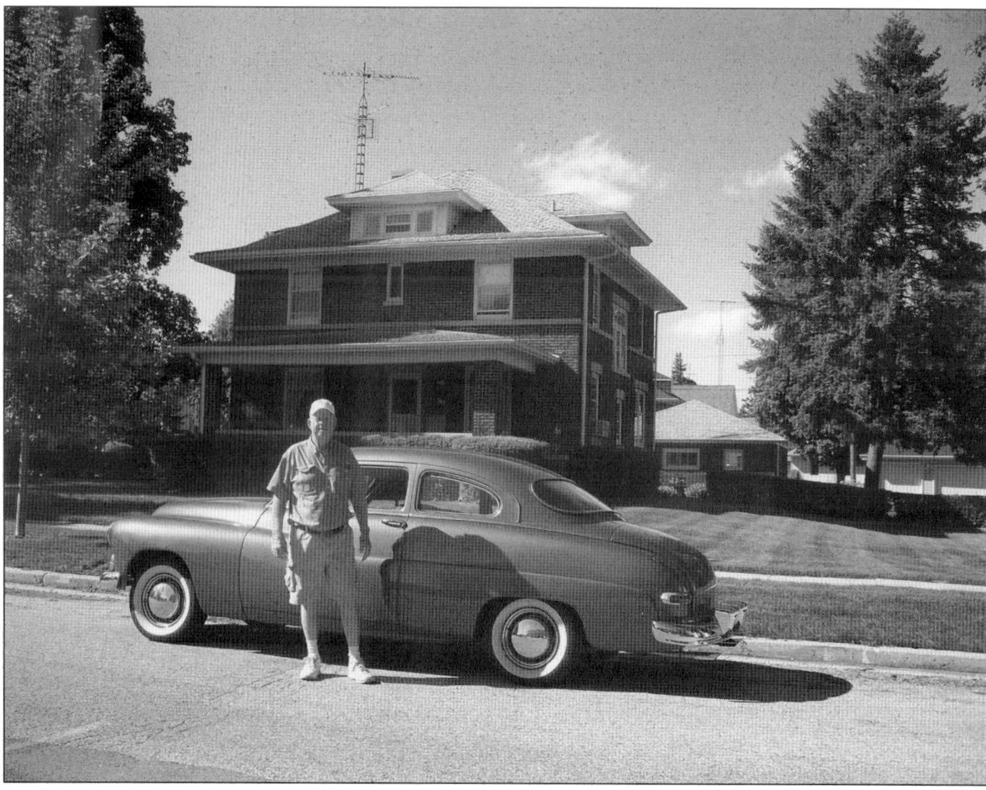

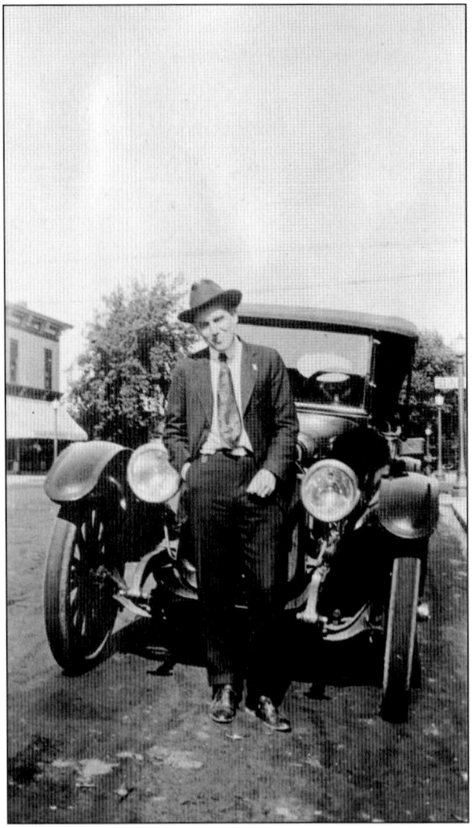

The picture is of Dr. Bowman, a dentist in Shabbona in the late 1920s, was given to the museum by Helen Pickwell. He is standing beside his Ford Model A, a model that came out from 1927 to 1931. It was the second huge success for the Ford Motor Company after its predecessor, the Model T. Prices for this car ranged from $385 for a roadster to $1,400 for a town car.

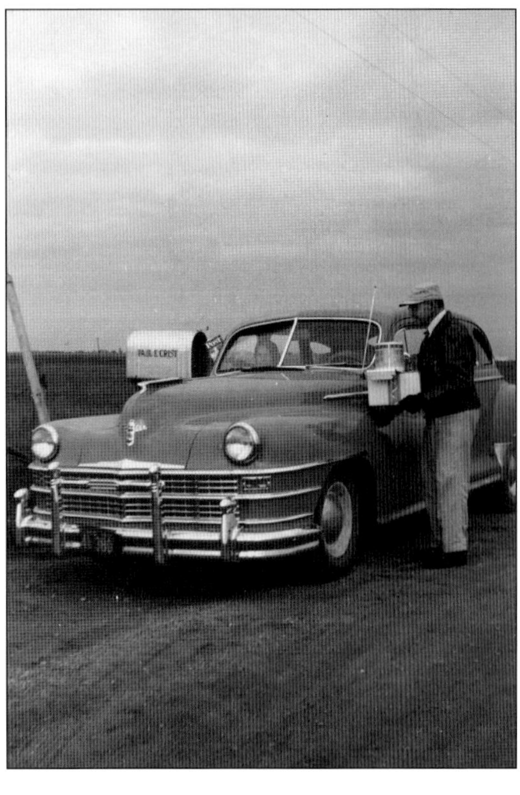

Floyd Olmstead was a rural mail carrier in the late 1920s to early 1930s. He is standing beside his delivery car, possibly an early model Pontiac. With the mail service paying an allowance for maintenance, rural carriers were responsible for furnishing all vehicle equipment. Note the woman in the car to help him. The name on the mailbox is Paul E. Crist, who lived on University Road in Shabbona, Illinois.

Five

COMMUNITY SCHOOLS

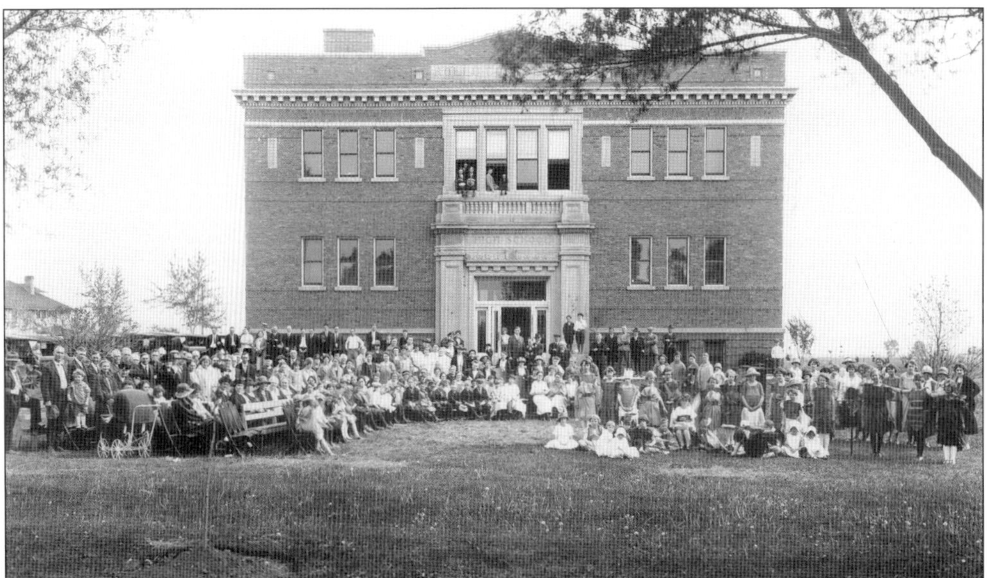

As enrollment declined in the one-room schools, districts voted in favor of consolidation. "One of the best examples in the United States of how rural schools may benefit from consolidation is the new township educational plant in Rollo, IL," stated *Collier's* magazine. On May 18, 1912, the voters decided to build a new school at Rollo, near where the one-room school was at the time. It opened on January 13, 1913.

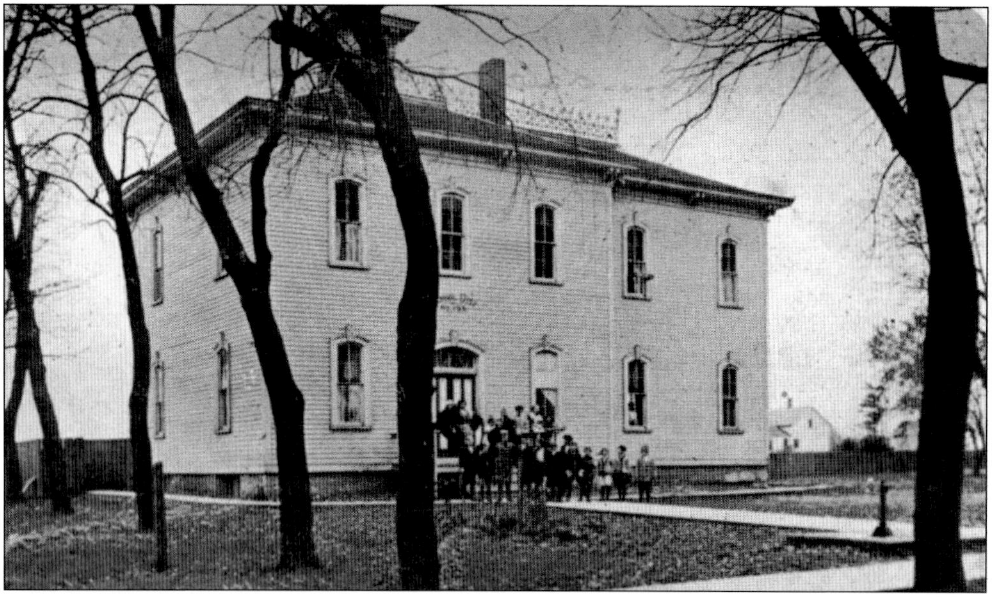

The Shabbona School District No. 125 was a two-story frame building erected in 1870 at a cost of $3,000. G.G. Knell donated land, and John Peter Swanson was the architect. There was a two-room addition in 1882 that cost $1,200, and a fourth room was added in 1886, housing eight grades. A three-year high school curriculum was added in 1887. The frame building was torn down in 1919.

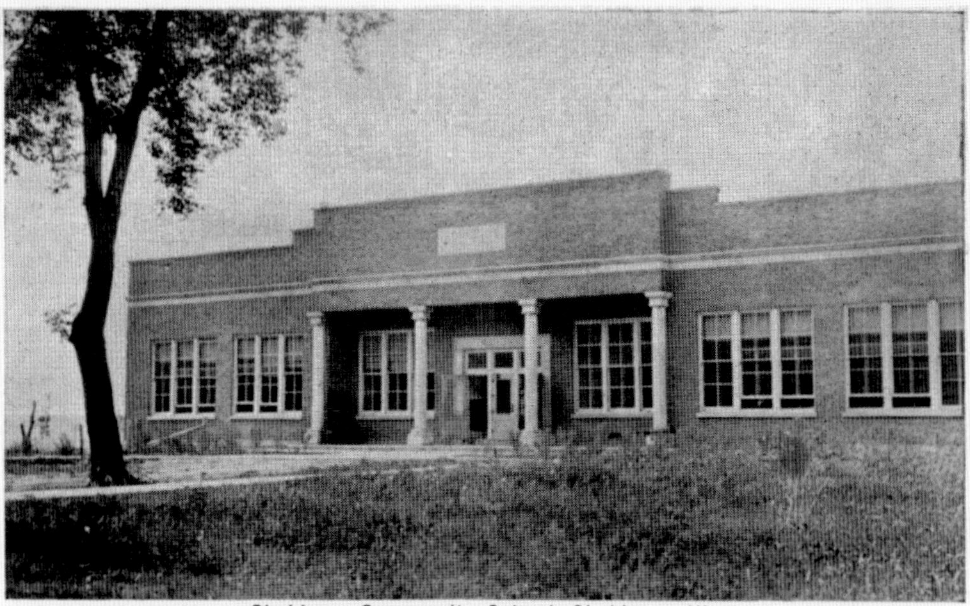

Shabbona Community School, Shabbona, Ill.

The Shabbona High School District No. 410 building, located near the site of the old grade school, burned in 1922 after only two years in existence. Principal and teacher A.J. Ladd, from the University of North Dakota, had put together one of the finest libraries in the state, and it went down with the building. On the following Wednesday, school opened in all departments in various places about town.

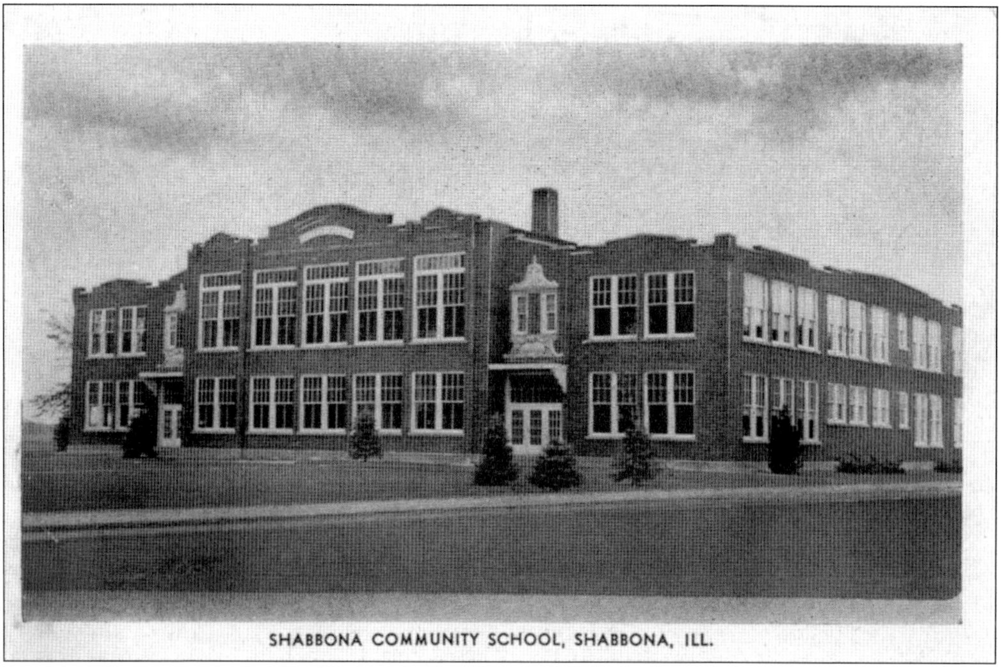

In January 1923, just 11 months after the school building burned, a new one was dedicated for Shabbona schools, uniting grades kindergarten through 12. The only change to the building over the years was the installation of the glass blocks, which replaced about three-fourths of each window. Community Unit District No. 425 was formed in 1949. The last high school students to graduate from this building were in June 1954; the building has since been torn down.

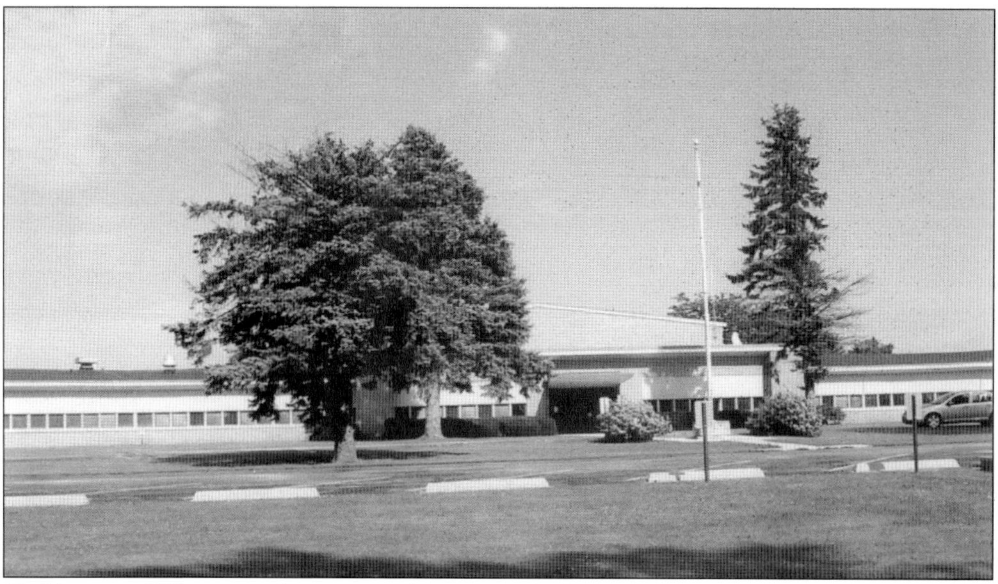

Shabbona and Waterman districts met for a series of talks over consolidation. In 1993, Waterman joined the Shabbona School District, and together, they became the Indian Creek School District No. 425. The Indian Creek High School is located on Shabbona Road in Shabbona. The Indian Creek Middle School is located in Waterman. There are two grade schools in the district—one in Shabbona and one in Waterman.

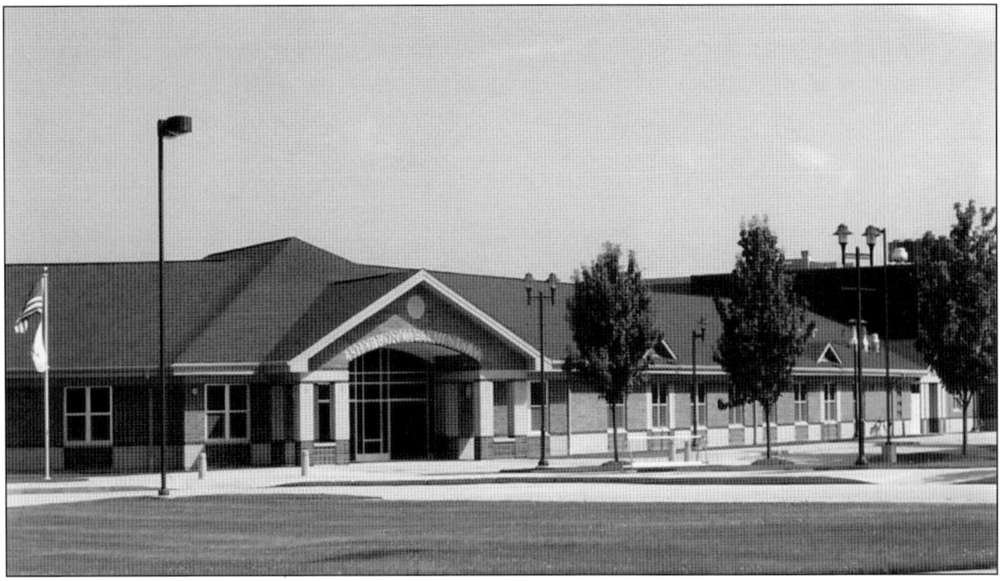

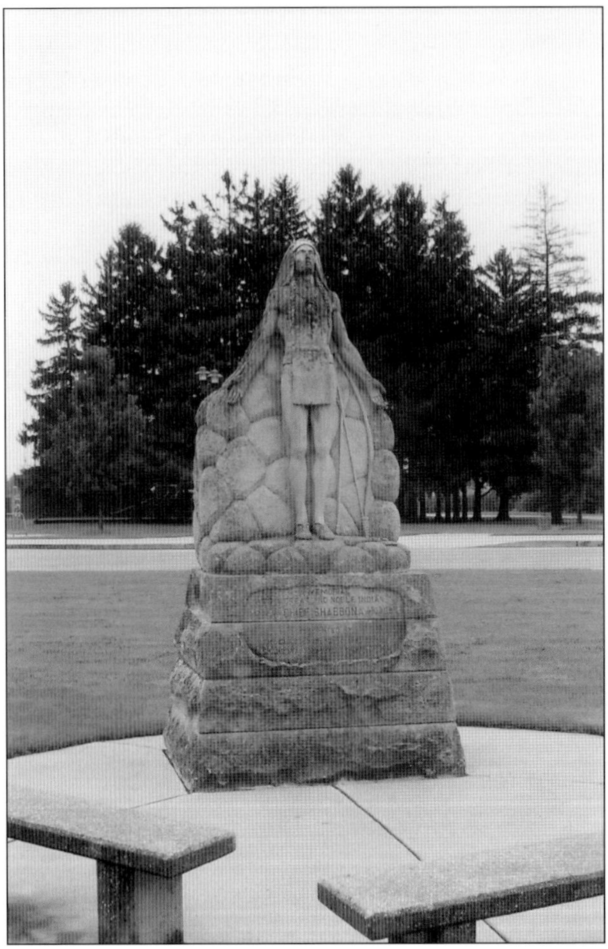

The new Shabbona Elementary School on Cherokee Street in Shabbona became a reality when the voters approved a referendum in 2003. The former 80-year-old building came down, and a new one in its place was ready for school in the fall of 2004. The 43,000-square-foot building housing 250 to 300 students ranging from kindergarten to the fifth grade has modern classrooms, a media center, a cafeteria, and a gymnasium.

On the grounds of the new Shabbona Elementary School stands the statue of Chief Shabbona, which was purchased by the classes of 1942, 1943, and 1944. Joann Pardridge's fifth graders put together and buried a time capsule at the base of the statue in the spring of 2005. It included pictures of students, newsletters, a calendar, a yearbook, a timber wolf Beanie Baby, pictures of old and new buildings, M&Ms, and other items.

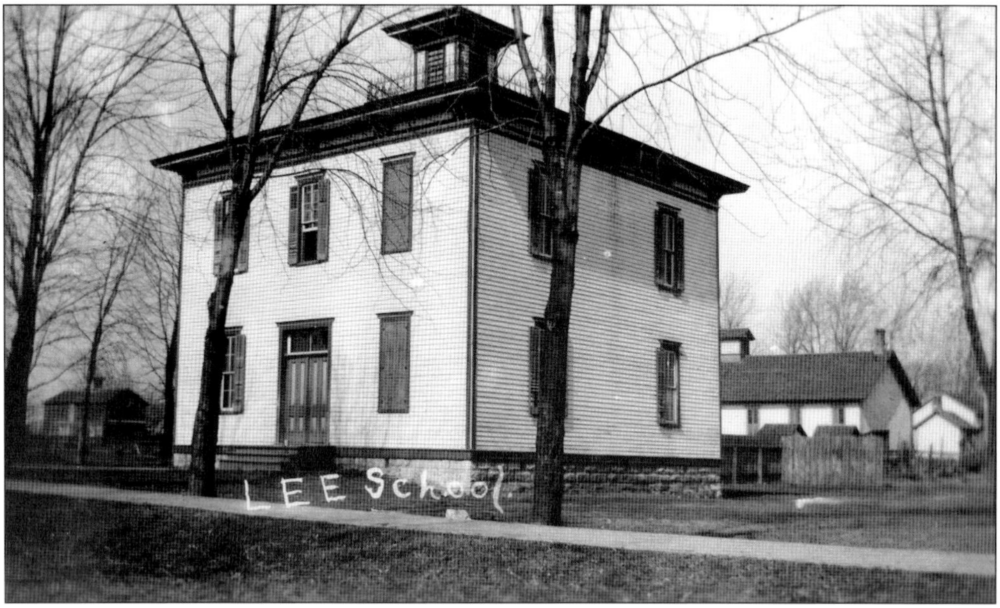

Religion and education were important to the early settlers. Parochial classes were held in homes. These classes were offered in addition to public school classes. A converted saloon was the first regular public school building. In 1874, the school within the village of Lee was held in the Dyas Building, with William Emmet as the teacher. He was the Sabbath teacher on Sundays in this building. A school was built in 1876, with Miss Hampton as the teacher for the first term. The Norwegians hired Thomas Nelson as the teacher of the Norwegian school. A larger frame building was erected in 1890, and the three-year high school addition came in 1918. Shown below is a Lee High School couple taking a drive in a new car. The picture was taken south of Lee, along Route 30 before the road was paved.

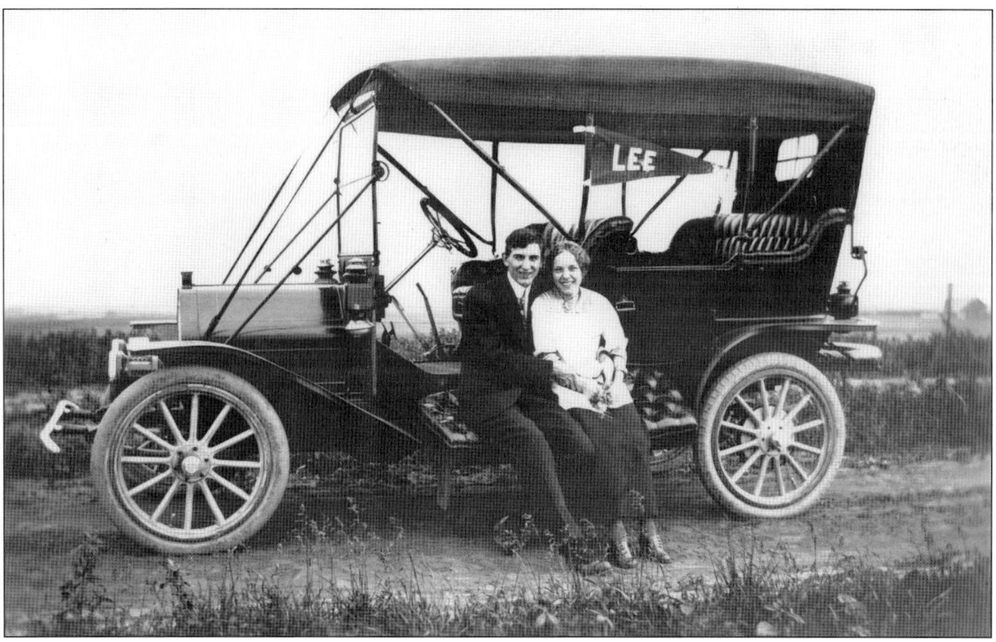

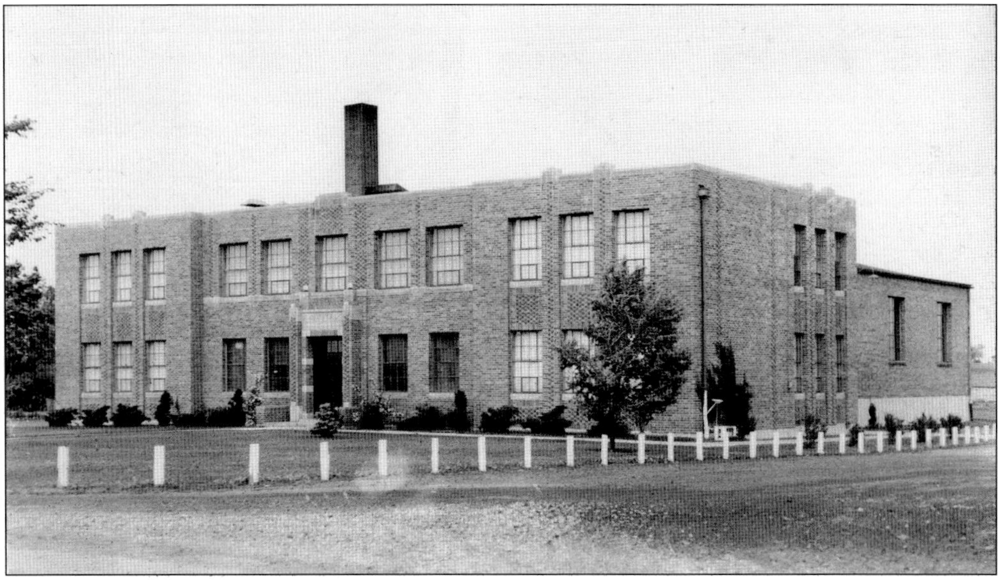

The Lee High School/Grade School was completed in 1936 on a 10.5-acre site, with Opperman and Sons of Pontiac, Illinois, as general contractors. The building was a two-story structure of yellow variegated brick with New Bedford limestone trimming. On the first floor are the two offices of the principal, a full-equipped foods laboratory, and the boys' and girls' cloak and lavatory rooms. On the top floor was the high school study hall and library with room for 65 students. On the second floor were four classrooms for high school use, including a science laboratory. A wing for elementary students was added in 1961. The school closed in the 1980s, and children were bused to Shabbona. Principal Burdette Peterson and his sixth-, seventh-, and eighth-grade classes in 1950 are shown in the picture below. (Below, courtesy of Nancy Fleming Pardridge.)

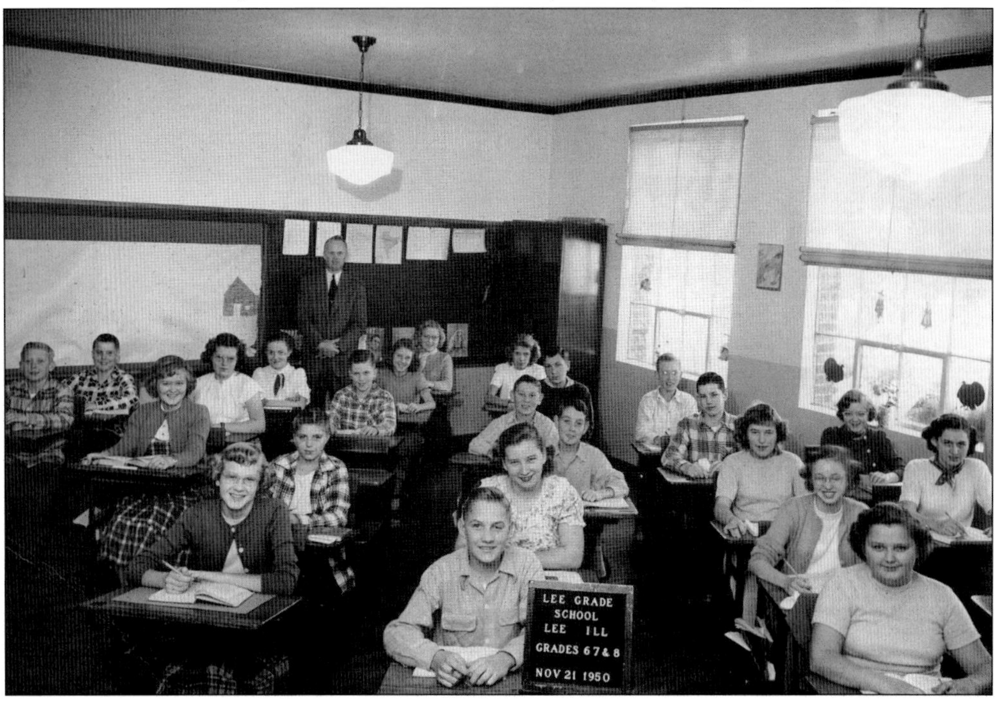

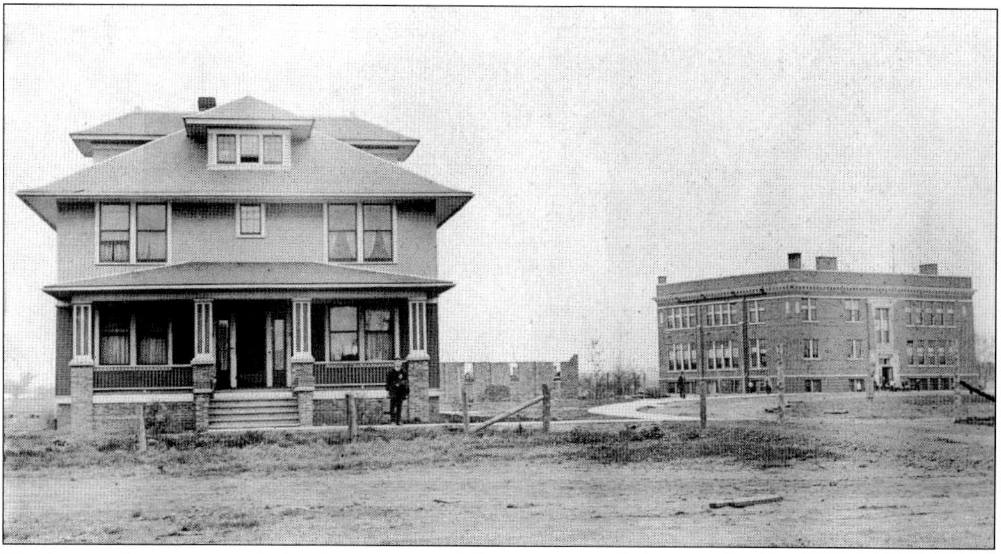

Playing an important role in the success of the new consolidation of schools at Rollo was the Teacher's Cottage. It was built in 1912 by the Olson Brothers of Shabbona on the 10-acre school campus at a cost of $6,500. The building burned to the ground in 1926. The outline of the foundation can still be seen today, in the left field area of the ball diamond.

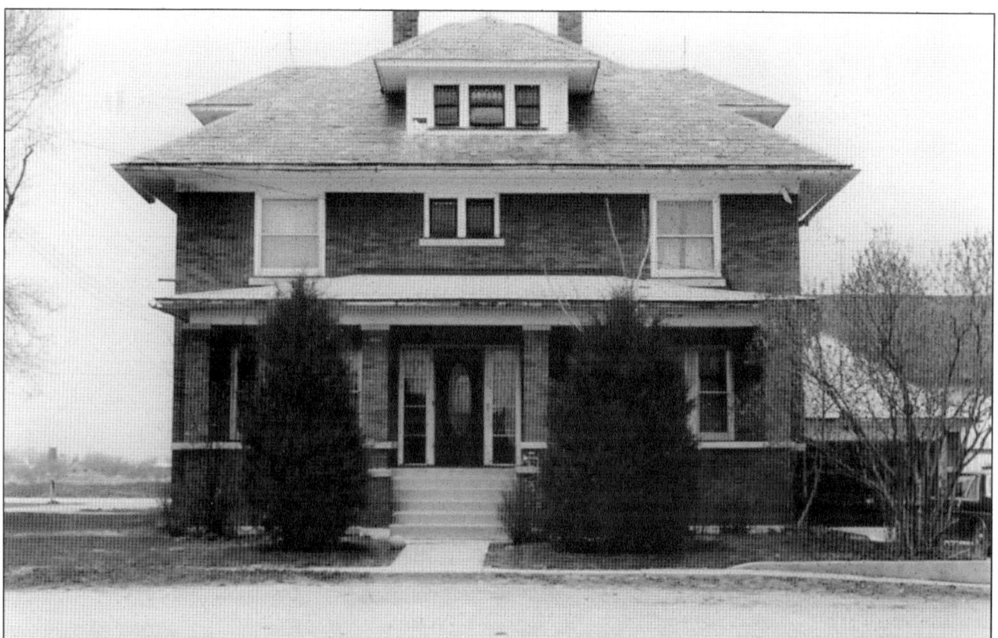

The board of education voted to buy the George S. Hyde house, built in 1914, for $8,000 for the teachers of the Rollo District. This was the large, brick, two-story residence located just west of the Congregational Church in Rollo. Farther to the north and entirely detached was a roomy barn and cement-block icehouse. After selling their house, Hyde family members moved to DeKalb.

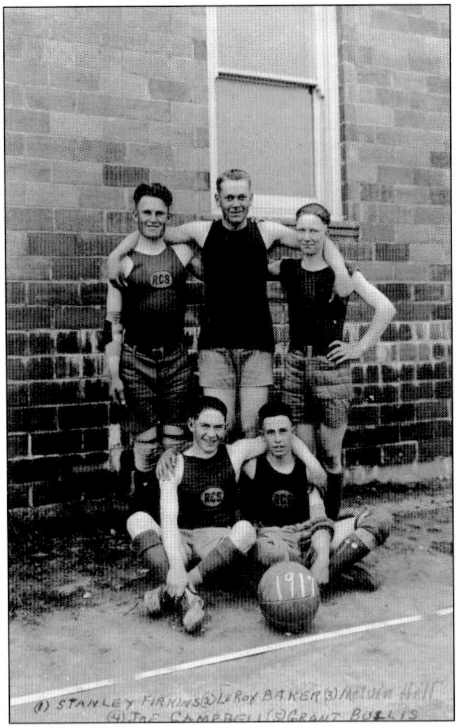

There might not have been a great many sports to select from at Rollo High School, but Rollo fielded teams in several sports each year. In 1917, the high school team played games outdoors because the high school gymnasium had not yet been built. Those pictured are, from left to right, (seated) Joe Campbell and Grant Bullis; (standing) Stanley Firkins, LeRoy Baker, and Melvin Hill. The picture was given to the museum by Doris Firkins.

Shown is a late-1920s or early-1930s Lee High School basketball team. From left to right are (first row) Orlo Edwards, Daniel Kennedy, Howard Mullins, and Seward Brown; (second row) Lewis Prestegaard, Oden Jordal, coach Tillman Knutson, Donald Herrmann, and Howard Fleming. The team won the conference for their high school. The picture was given to the museum by Helen Pickwell.

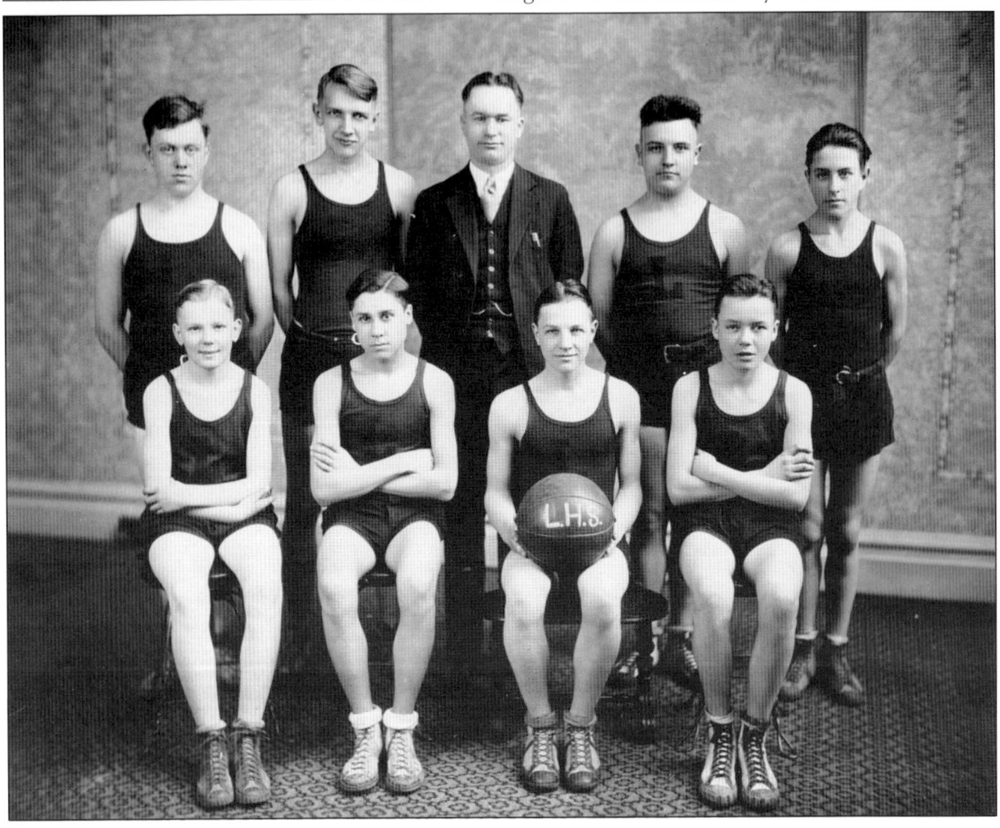

Hiram Johnson formed a women's softball team known as the Farm Bureau girls' softball team in 1934. It was made up of Shabbona and Waterman girls who were, at most, four years apart. This team was known for their hard work, good sportsmanship, and stamina. The home games took place at the Shabbona Grade School. In 1937, the team played the Peoria County Farm Bureau team for the state championship and won. Those pictured are, from left to right, (first row) Jean Dugan and Martha Mullins; (second row) Sally Spangler, Ruth Challand, Edna George, and Lorraine Colby; (third row) Lois Challand, Dorothy Degan, Lucille Larson, Hiram Johnson, Max Conway, Altha Thuestad, and Helen Challand. The picture below shows a 1926–1927 men's baseball team from Shabbona. Among the players pictured are Amos Oftedal, Wilbur Challand, Ted Watne, Harold Carson, Ed George, Kenneth Storey, Gib Hallam, and Kenneth Flewellin. The photograph to the right was given to the museum by Martha Mullins and Lucille Larson.

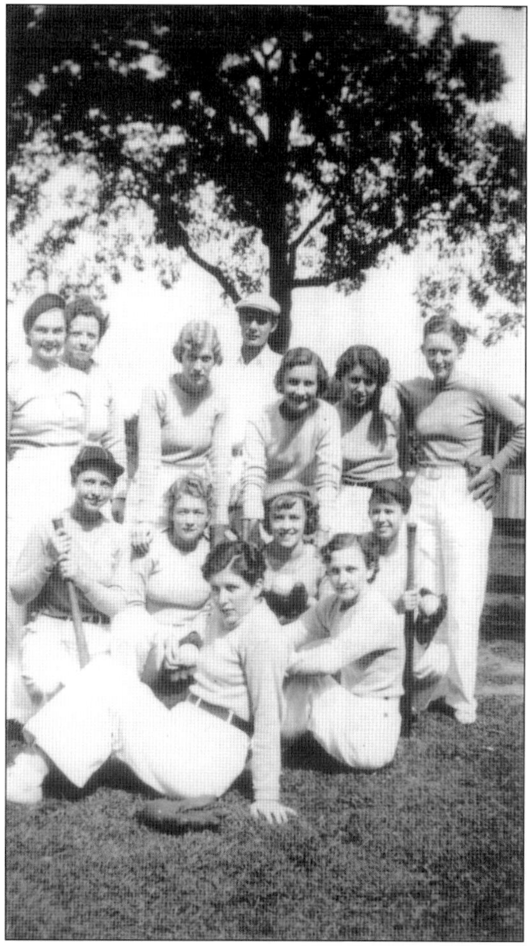

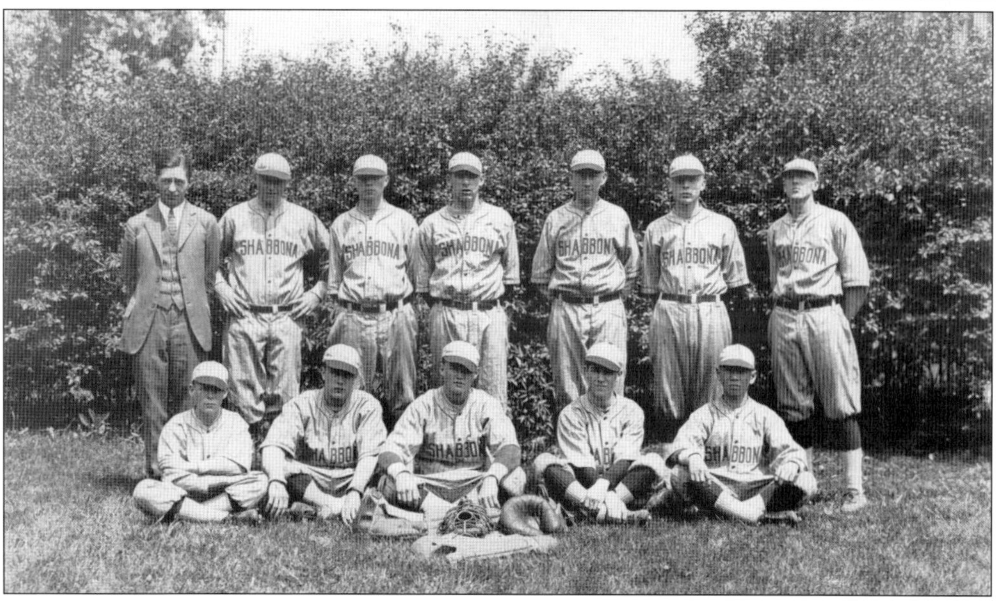

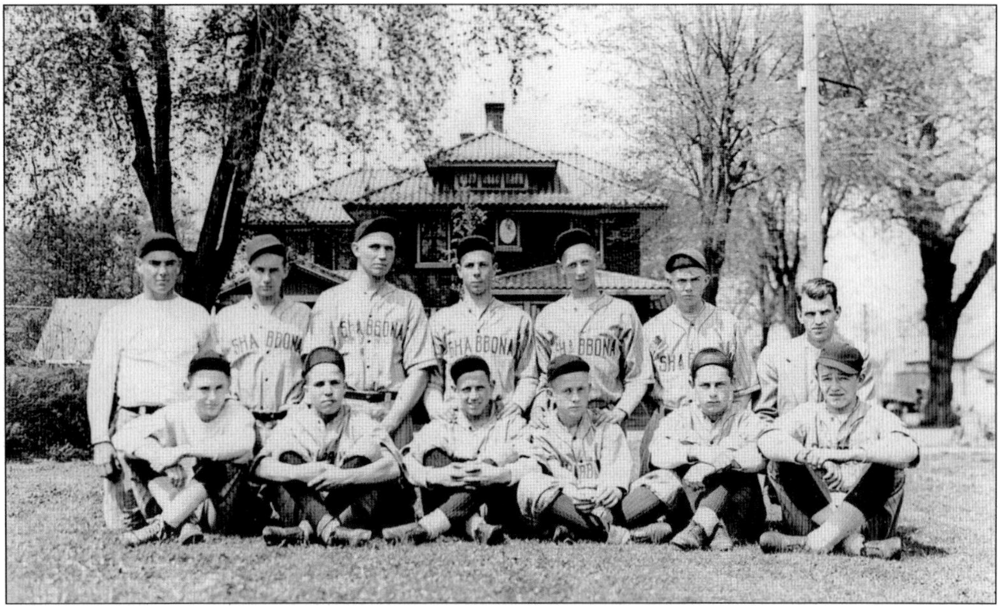

The Shabbona High School Future Farmers of America softball team of 1936 was the only one in Shabbona to win a state championship. Among the players were Coach Matthews, Harry Houghtby, Harold Cox, Bob Hickey, Joe Craig, Gilbert George, Bud Challand, Paul Oleson, Bernard Schnorr, and Ray Schoenholz. The trophy is showcased at the Shabbona-Lee-Rollo Historical Museum. (Courtesy of Jim Cox.)

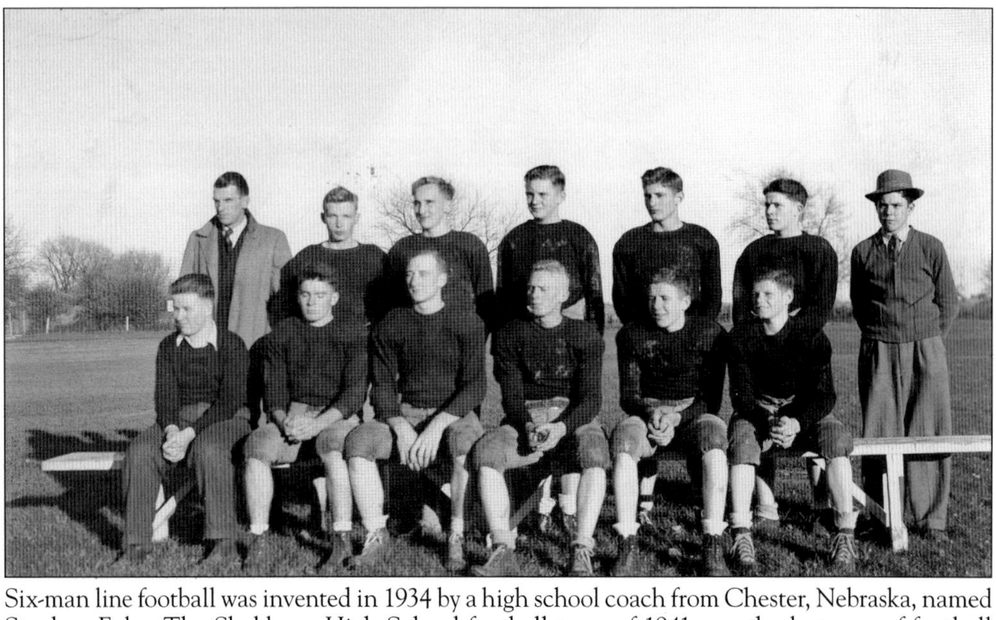

Six-man line football was invented in 1934 by a high school coach from Chester, Nebraska, named Stephen Epler. The Shabbona High School football team of 1941 was the last year of football (a six-man team) at Shabbona. Players included Dean Snow, Paul Donnelly, Milton Durr, Bob Jacobson, Bob Johnson, Leslie Skinner, coach Roy Guttshaw, George Lovering, Alvin Watne, Stanley Houghtby, Bernard Herrmann, and Bill Smith. (Courtesy of Bob Johnson.)

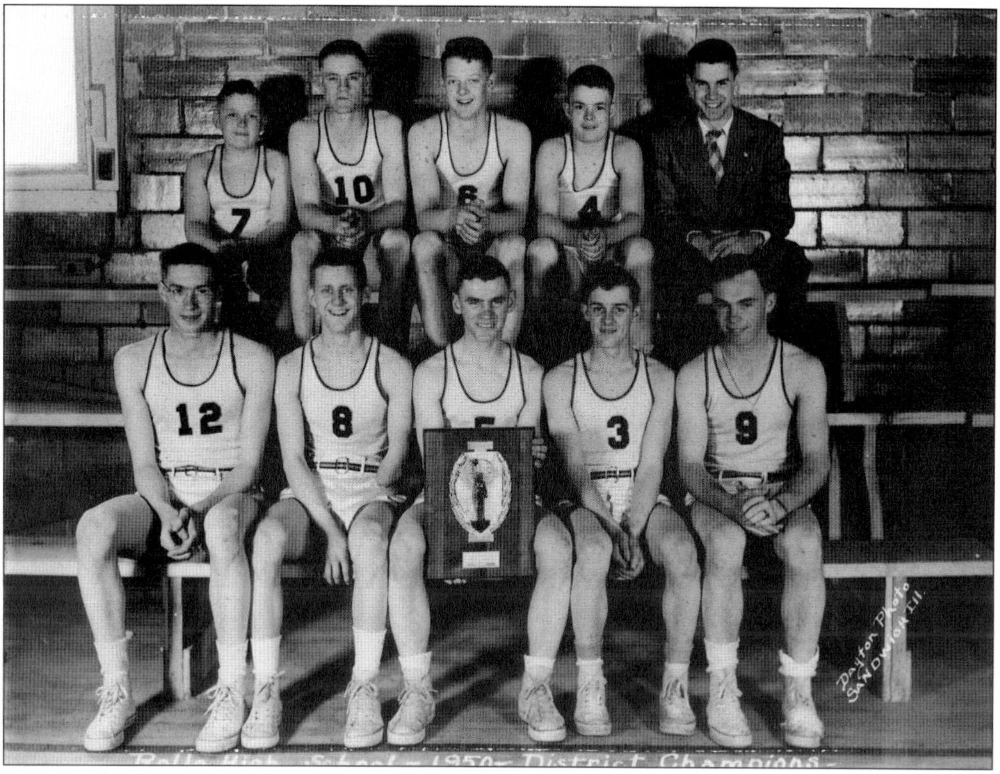

Rollo High School had an enrollment of 23 students in the 1949–1950 school year. Ten were boys. Nine of those boys made up the district tournament team. The team won the district and lost in the regional to Marseilles by a score of 49-47. Seen here are, from left to right, (first row) Phil Buland, Alvin Benson, Dan Todd, Wayne Goldsberry, and Robert Landers; (second row) George Durbin, Robert Catron, Gilbert Iverson, Donald Landers, and coach Carson DeJarnatt.

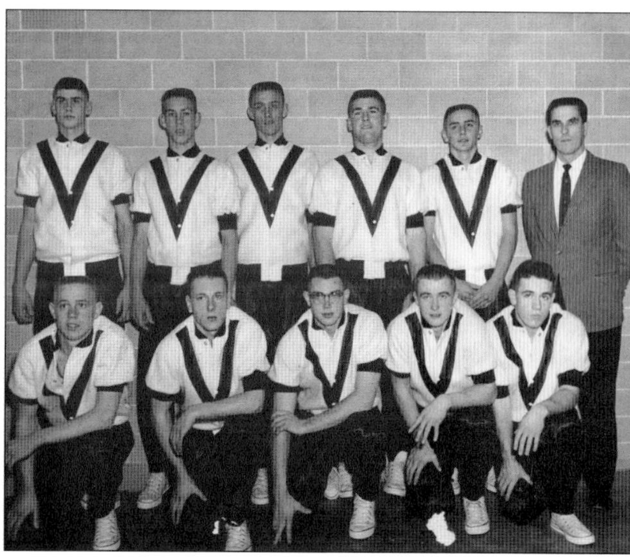

The Shabbona High School 1958–1959 basketball team won 32 and lost 1. Members were conference champs, conference tournament champs, DeKalb Holiday Tournament champs, and district champs. The team also won the regional tournament. They lost in the sectional to Hinsdale. Shown are, from left to right, (first row) Charles Svendsen, Ken Flewellin, Bill Mullins, Denny Fleming, and Ron Hart; (second row) Marv Straw, Dave George, Ron Espe, Jerry Wisted, Rich Benson, and coach Francis Wagner.

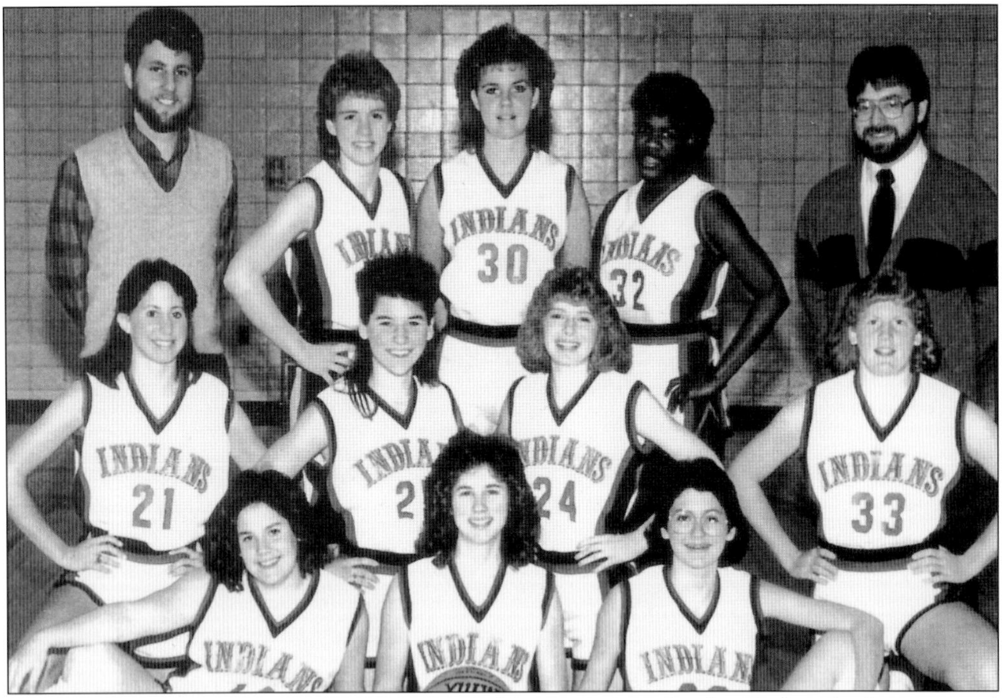

The 1986–1987 girls' basketball team at Shabbona High School ended with the sectional championship game. The girls finished the year winning 22 games—including conference champs, Little Ten Tourney champs, and regional title—and losing 2. From left to right are (first row) Joni Barshinger, Dorothea Hart, and Lisa Lytle; (second row) Chris Wallace, Jeanette Swatkowski, Sherry Dunn, and Julie Thompson; (third row) coach Ken Dawson, Kristen George, Amy Todd, DeeDee Jeske, and coach George Limberis.

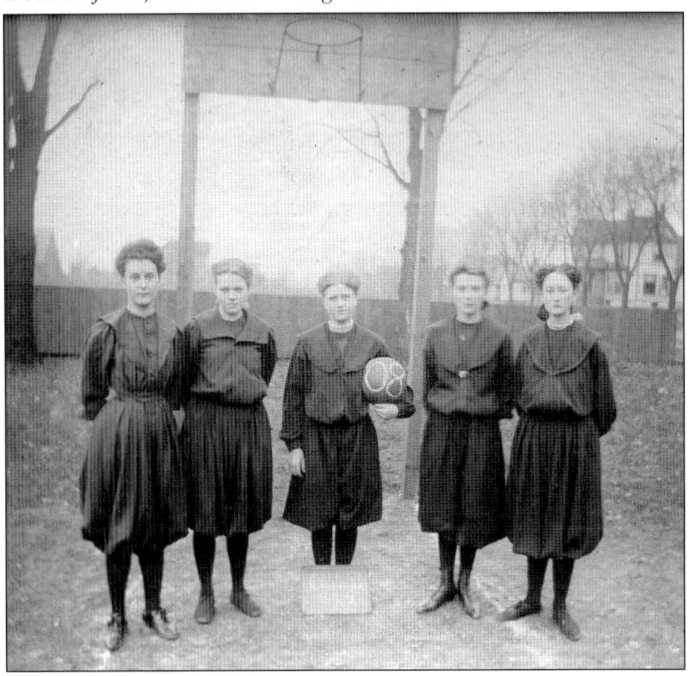

Basketball is one of the few women's sports that developed in tandem with its men's counterpart. It became popular in large part because of women's colleges. The picture is of the Shabbona High School women's basketball team of 1908. Players are Ferrol Grant, Deca Ostewig, Mary Holbo, Alice Thompson, and Nina White. Notice the women playing in dresses and on an outdoor, grass court.

Six

People and Places

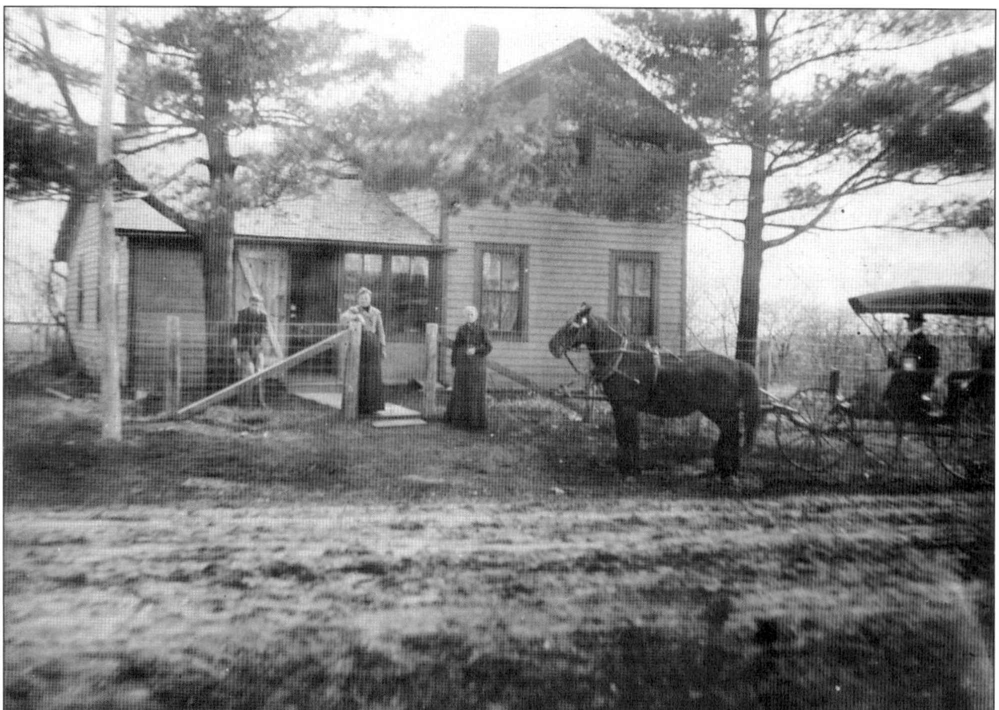

In 1834, Paw Paw Township in DeKalb County became the home of the white man when David Towne settled on the southwest quarter of Section 19. David Towne arrived in the autumn and did little that winter—except build a cabin and hunt. The Montolbert Greenfield home built in 1854 was located in Shabbona Grove across from the schoolhouse. Lumber for the house was brought from Aurora by oxen.

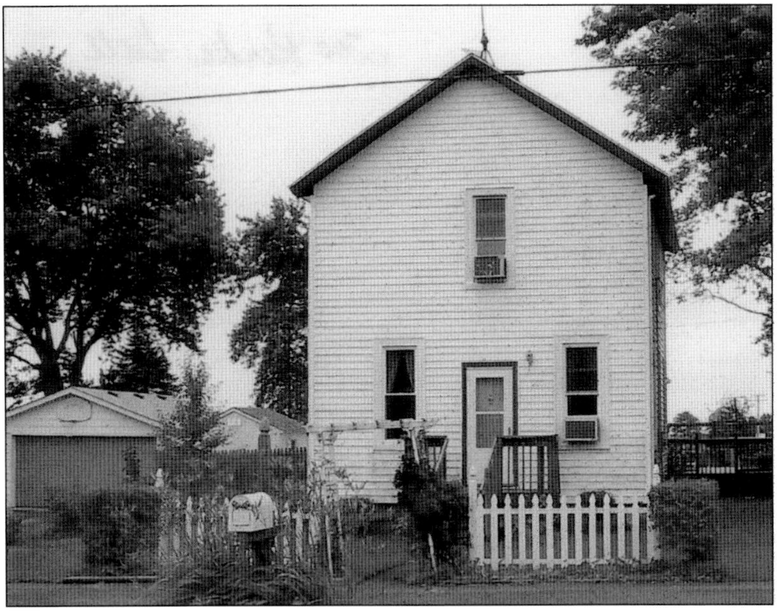

The home of Joseph and Tina Phipps at 200 Kirke Gate in Lee is the oldest house on the DeKalb County side of town. Christopher Christopher came to America in 1855 and built this house in 1874. He and Jorgens Prestegaard erected the first building in Lee, a small grain office and grocery, and put up the south elevator in 1873.

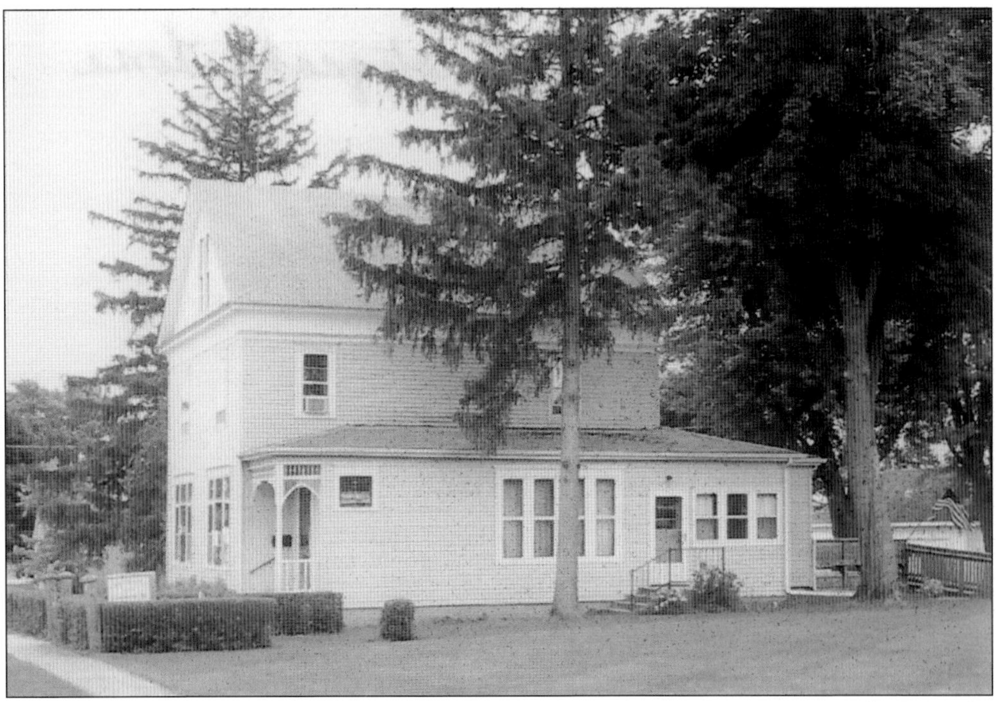

Mary Ellen Prestegaard's home on the Lee County side of the town of Lee was built by the Youngren brothers and is one of the oldest houses in Lee. The Ole J. Prestegaard family moved into Lee in 1871 and engaged in the lumber, coal, and grain businesses for 16 years. He was a school director, a deacon, trustee of the Lutheran Church, and a stockholder and director of the Lee State Bank.

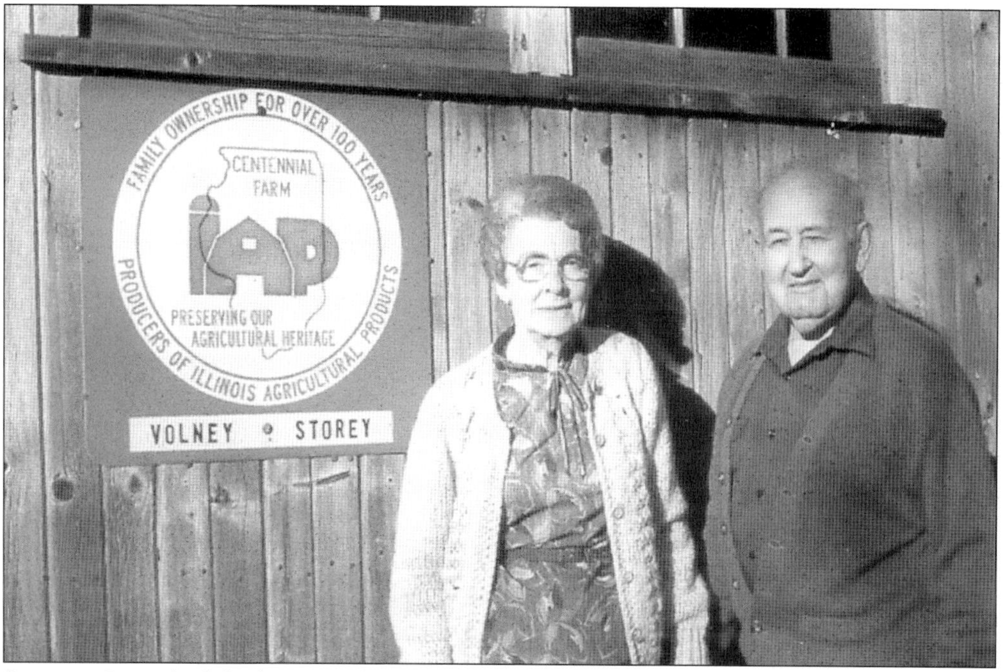

Septimus Storey built his homestead on Route 30, west of Shabbona, in 1856. One of nine children, his son Deforest married Jane Henning and bought 160 acres in 1907. Grandson Volney (shown with wife Edessa) farmed for 28 years until a heart problem made his retirement mandatory. The land is now co-owned by great-grandchildren—Nancy Joan Storey and Donald E. Storey. These family members continue to be involved in its management.

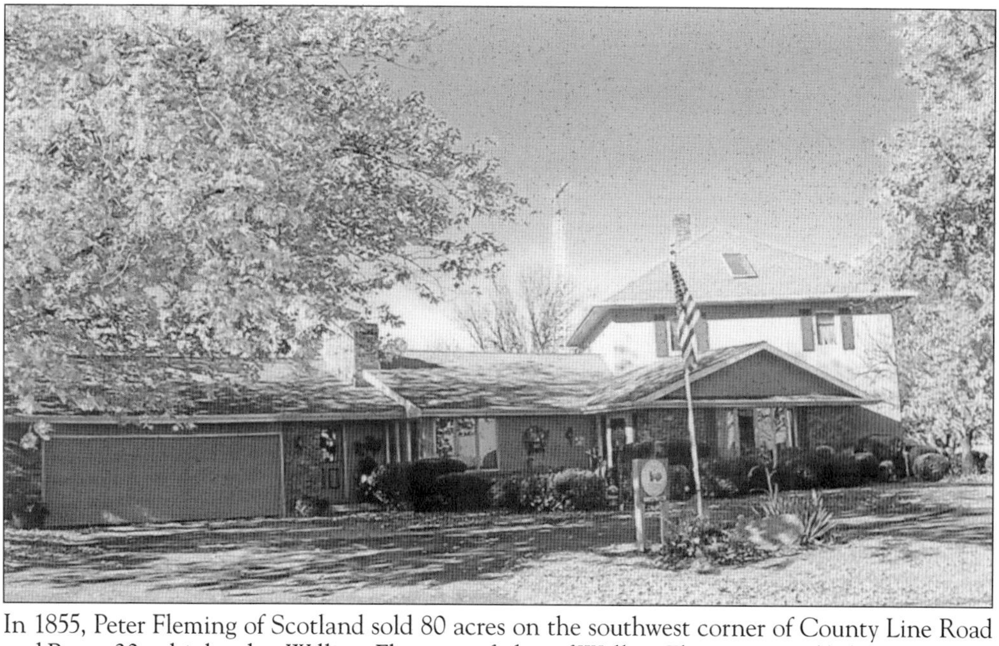

In 1855, Peter Fleming of Scotland sold 80 acres on the southwest corner of County Line Road and Route 30 to his brother William Fleming—father of Wallace Fleming, grandfather of Howard Fleming, and great-grandfather of Dennis Fleming. The home of Dennis and Marlene Fleming is on County Line Road in Lee, Illinois.

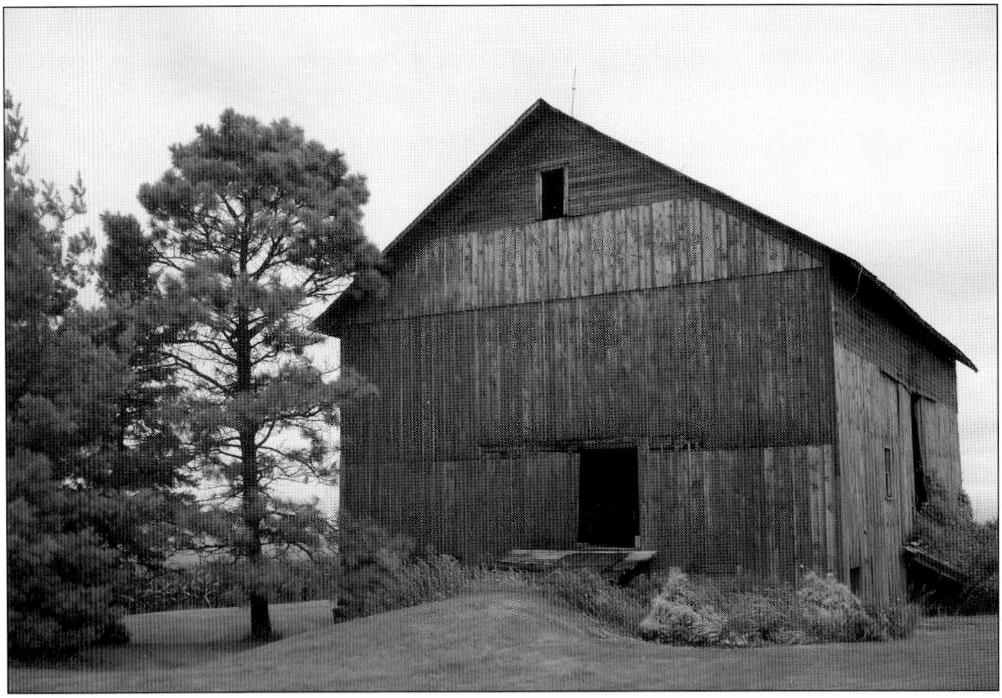

Asher Downer came to Milan Township in 1859. In time, he acquired 440 acres. Asher is said to be the first in the area to pump water for stock with a gasoline engine. Subsequent occupants were the A.W. Cofield, E.R. Cofield, and the Edwin Cofield families. The barn was built in 1882 and is still standing. The home of Heath and Julie Headley is on Tower Road, north of Shabbona.

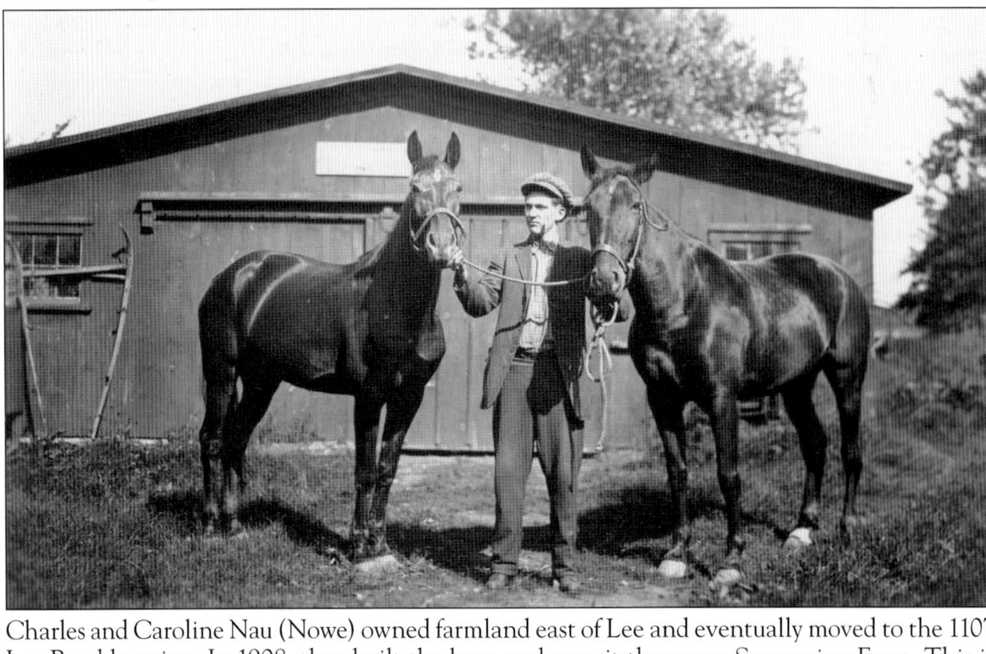

Charles and Caroline Nau (Nowe) owned farmland east of Lee and eventually moved to the 1107 Lee Road location. In 1908, they built the barn and gave it the name Sunnyview Farm. This is the home of Jack and Anne Flewellyn Hart that was north of Shabbona. The picture is of Wayne Flewellyn, the father of Anne, with his horses.

Theodore O. Berg came to America from Norway in 1853, married Maria Danielson in 1858, and had five children: Elizabeth, Caroline, John, Amelia, and Oscar Theodore Dedric. He helped organize the DeKalb Agricultural Association. He moved to the farm on Tower Road around 1900, calling it the Clover Leaf Stock Farm. It is the home of Roger and Kathleen Knutson (Kathleen is the great-granddaughter of O.T.D. Berg). (Courtesy of Dave Jordal.)

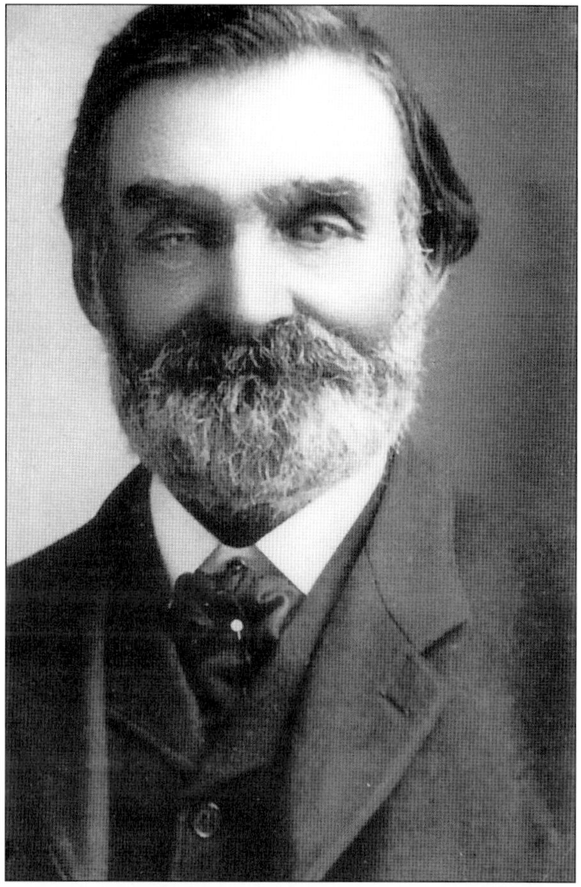

Nels Peterson, great-grandfather to present owner Charles Thorsen, bought 160 acres in 1882. He and his only son, Isaac, farmed the land on Woodlawn Road north of Lee. In 1841, Gladys and Orland Thorsen took over ownership of the homesite, a fourth-generation cattle and grain farm. It is the childhood home of current Illinois senator Christine Johnson.

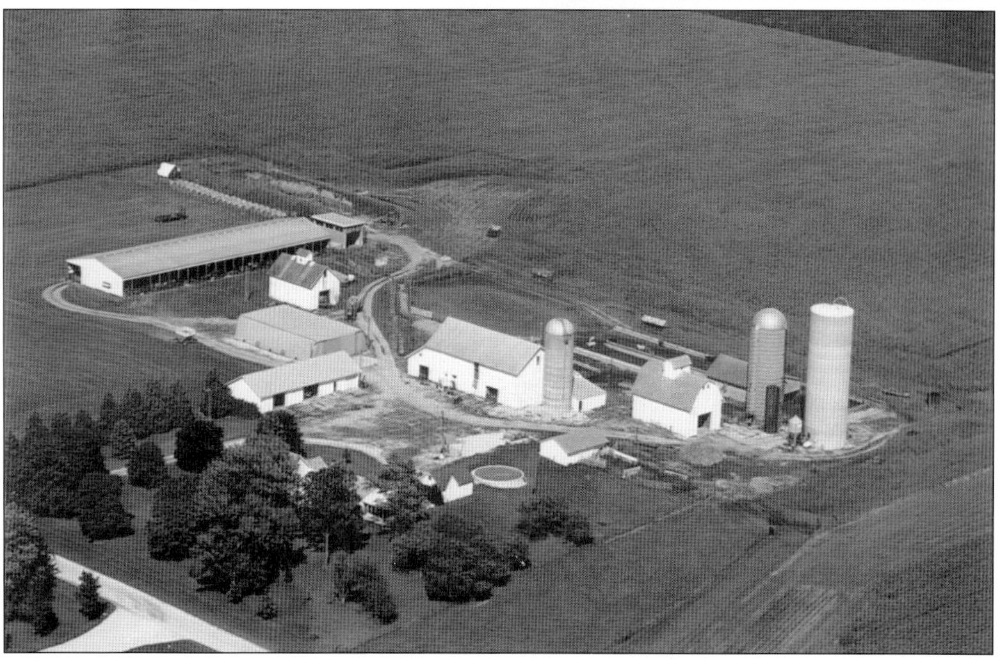

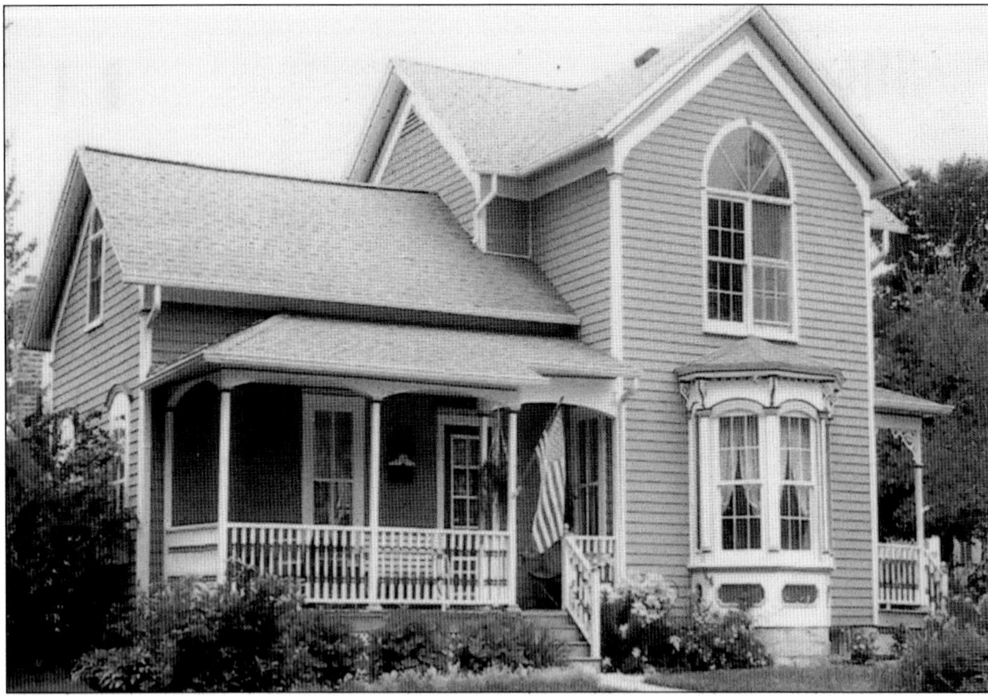

The home of Jared and Cary Jo Goodrich at 208 East Cherokee in Shabbona is the oldest house built on-site in 1872. Land was purchased from John Palm, and William Husk was the first owner. It originally was enclosed with a white picket fence. A barn was torn down and used to build the John and Marion Munson house across the street in 1938.

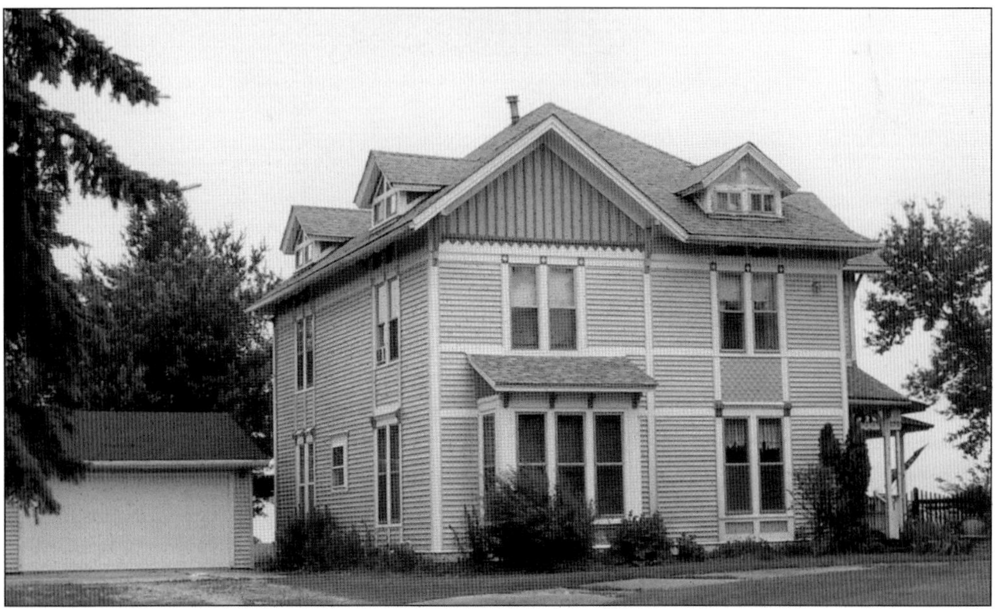

The home of Richard and Judy Mayer in Shabbona was originally the old Boslough home. It sat at the north end of the street along the east side of the Congregational Church (now Hiawatha Street). Due to the housing development on Hiawatha Street, the house was moved a short distance north and around the corner to Pueblo Street in 1977.

At age 21, Septimus Storey, born in England, came to the United States with a group of friends in September 1850. They eventually established a community west of Shabbona. Septimus was a member of the school board, road commissioner, and supervisor. Storey family members are, from left to right, (first row) Lillian, Septimus, Mary, and George; (second row) John, William, Sarah, Charles, and DeForest. (Courtesy of Nancy Storey.)

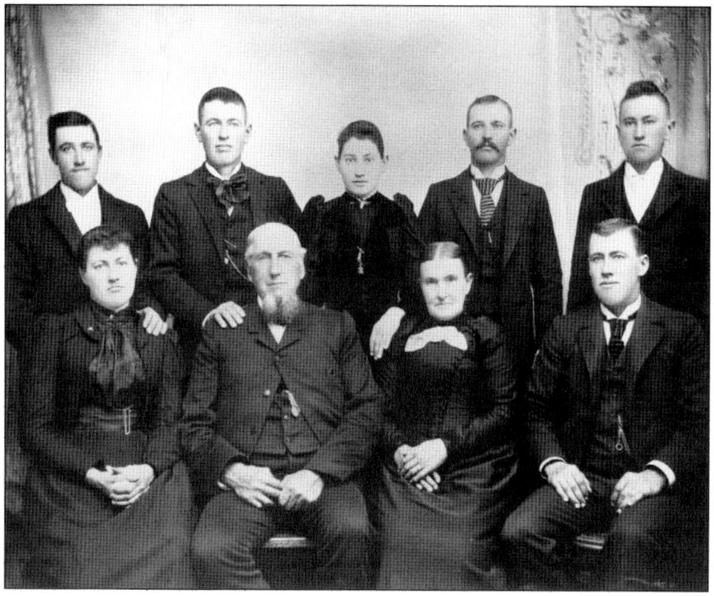

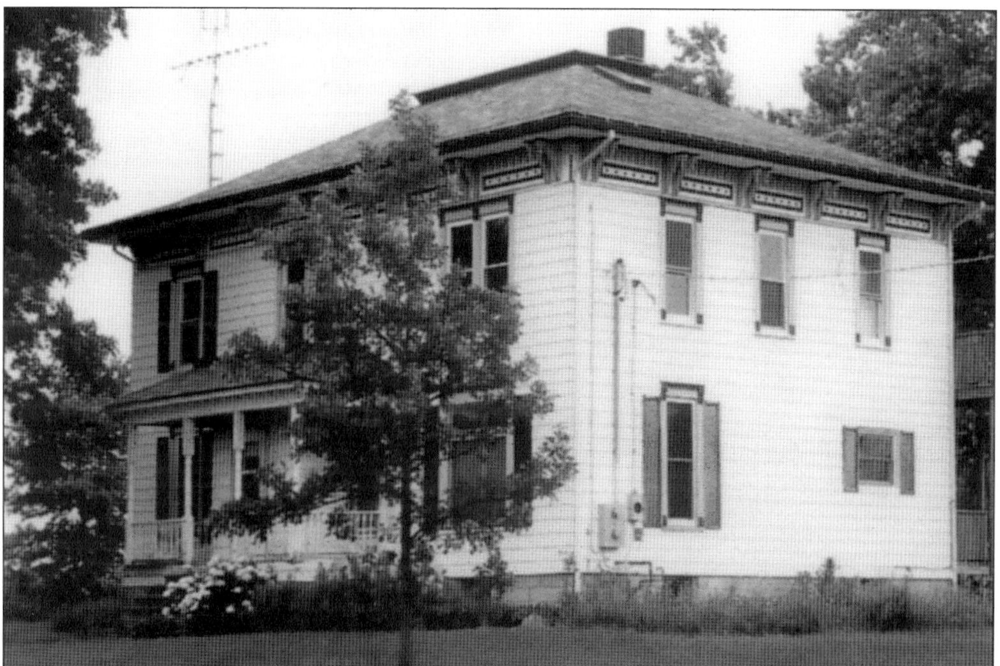

The home of Rudy and Karen Knappmeyer at 403 East Cherokee Street in Shabbona was built for Septimus Storey by William Swanson. It is an Italianate-style, two-story frame house built in 1895. One of the former owners found the Swanson name on the rafters of the house. A carpenter told her it must be a Swanson house because of the workmanship. He said they always used two nails where they only needed one.

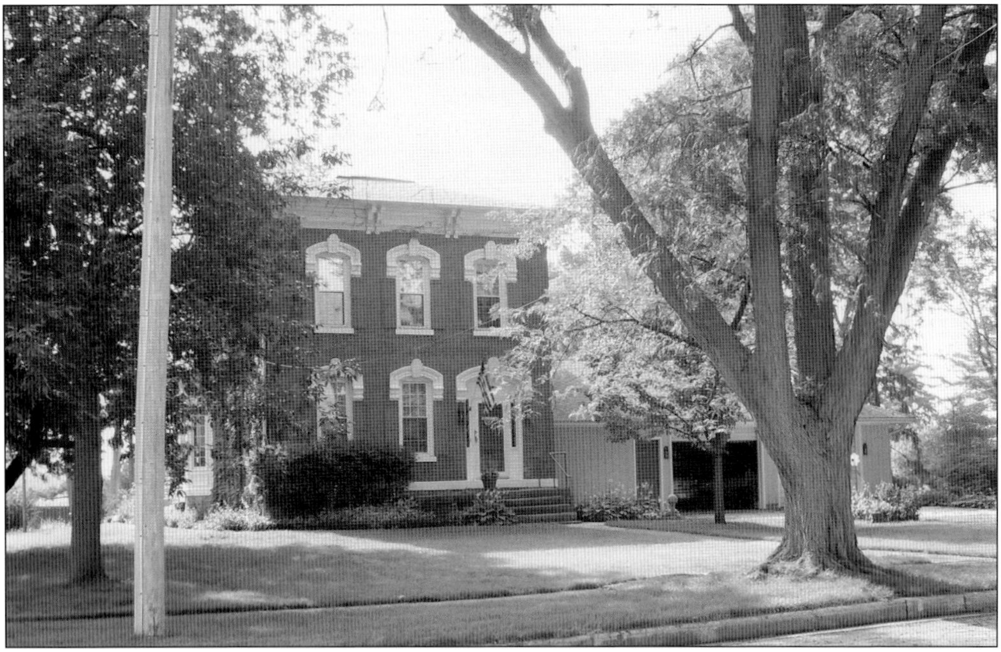

The Tom and Pat Malohn home at 213 North Illini Street in Shabbona is an example of a finely detailed two-story, Italianate-style structure that sits on a limestone foundation and has a characteristic low-pitch hip roof with wide overhanging eaves. It has paired brackets and frieze panels below it. The house was built in 1877 for Madison Van Velzor.

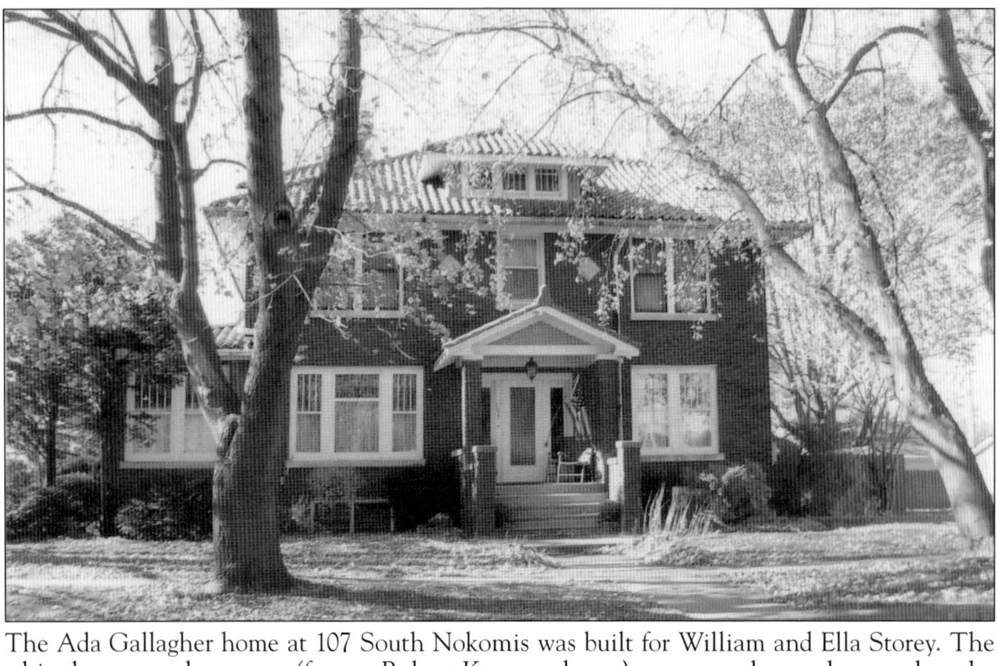

The Ada Gallagher home at 107 South Nokomis was built for William and Ella Storey. The white house on the corner (former Robert Knutson home) was moved next door so that the Federalist-style home could be built. The bricks are the same as the ones in the old Shabbona Grade School, and leaded-glass windows are throughout the house. Ada is the current curator of the Shabbona-Lee-Rollo Historical Museum.

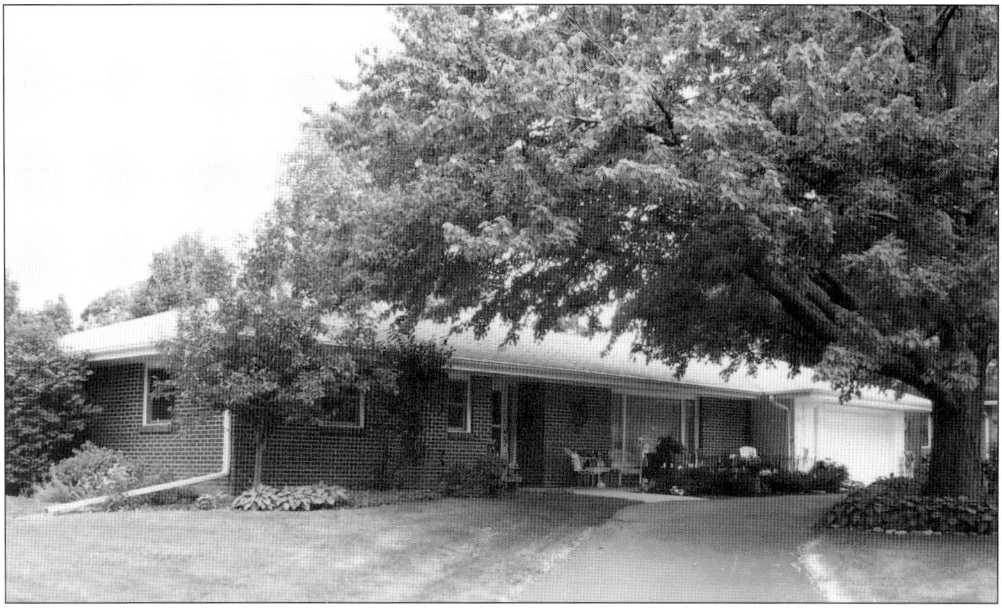

Lane and Hazel Clapsaddle (benefactors of the museum) built the brick Prairie-style home at 211 North Hiawatha Street in Shabbona in 1958. They moved from a farm south of town into Shabbona. Lane passed away in 1990, and Hazel continued to live in the house until she passed away in 1994. It is the home of Don and Nancy Pardridge, current president of the Shabbona-Lee-Rollo Historical Museum Board.

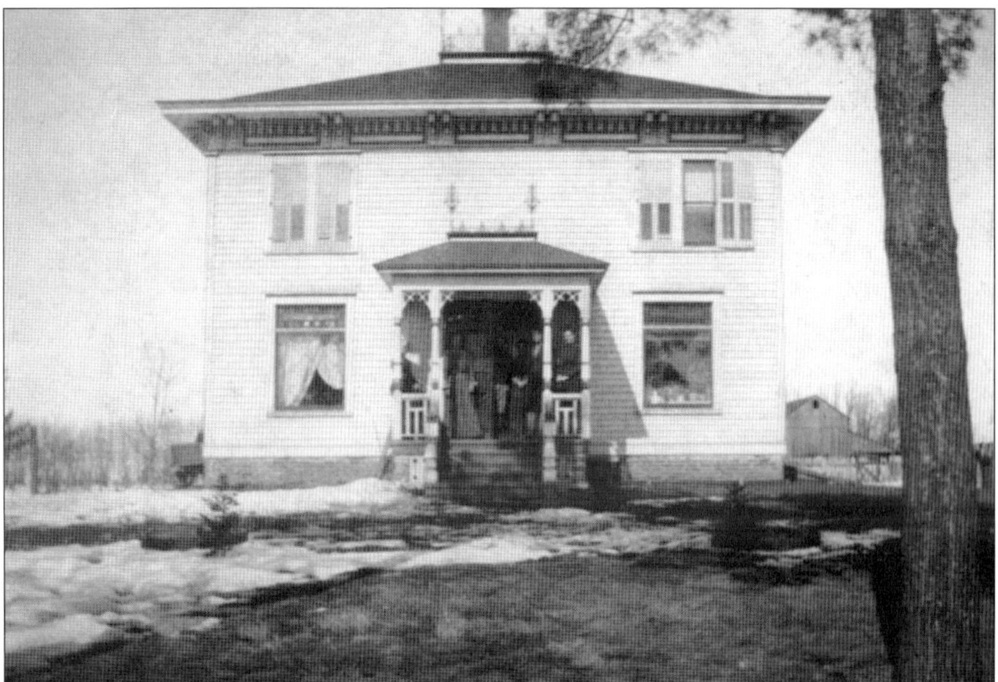

Named in honor of the family who has lived along its path the longest, Bend Road is a tribute to the Bend family heritage. David Bend is the fourth generation in his family to farm their plot of land in Shabbona Township. This farm is the home of David and Carol Bend.

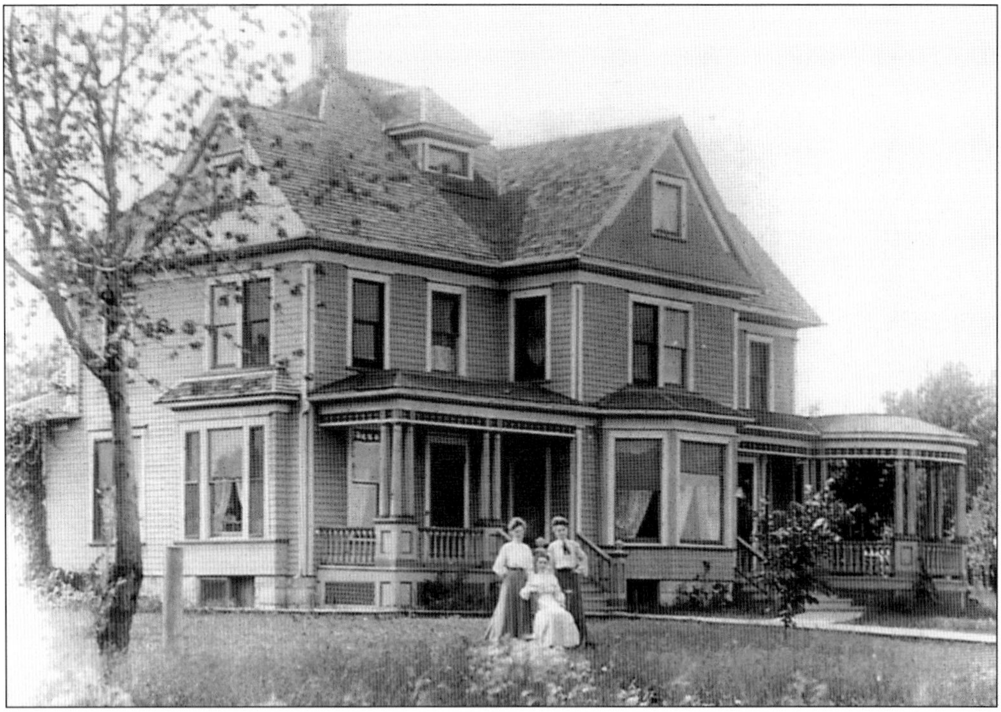

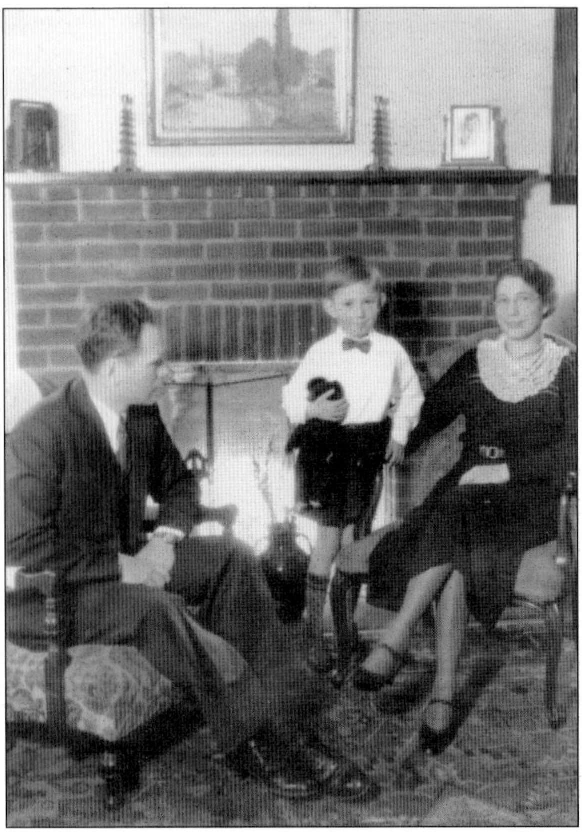

William Spray came to Shabbona from England in 1860, married Maude Kittle in 1905, and operated the local meat market. They purchased the house on Shabbona Road from Reuben Challand around 1913. William and Maude Spray and daughter Martha moved from the country on Shabbona Road to the town of Shabbona at 212 North Illini Street, where Martha married Howard Mullins. It is the home of Bill and Bonnie Mullins. (Courtesy of Bonnie Mullins.)

Donald C. Henderson Sr. grew up on the family farm purchased in 1884 in Victor Township. The family produced and exhibited Shropshire sheep, Poland China hogs, shorthorn cattle, and Percheron horses. He attended Rollo Consolidated High School, making the trip by horse and buggy and, later, by Model T Ford. One of his many accomplishments was to serve as a Fulbright lecturer in Tehran, Iran. (Courtesy of Donald Henderson.)

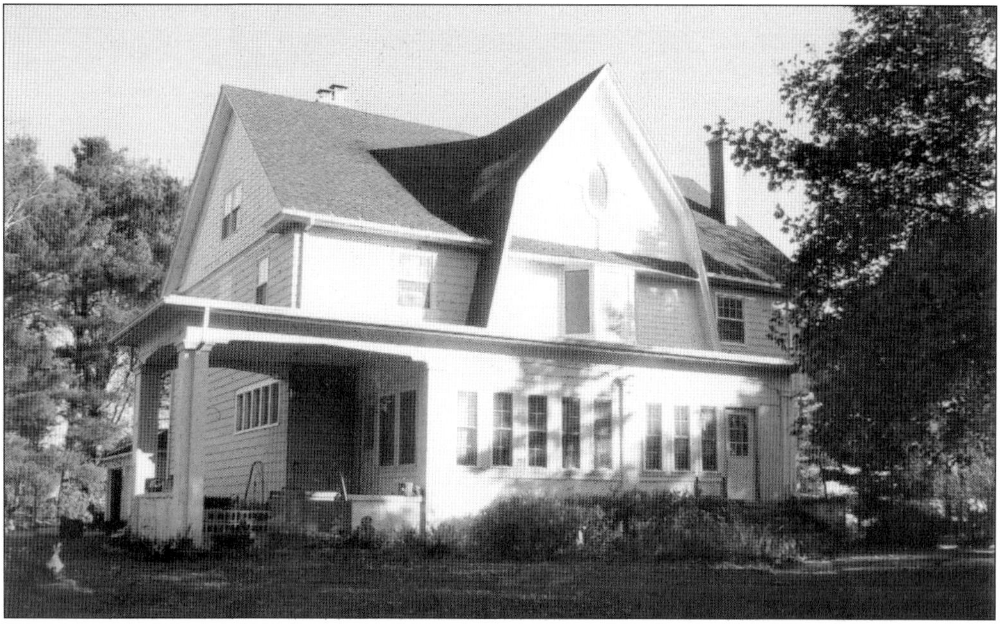

Printed in the September 1906 *Sycamore True Republican* newspaper was a write-up that read, "Finest Farm Home in DeKalb County, home of Dan J. Shepardson. It is a frame building of novel, but harmonious design." Several of the rooms are finished in black walnut, which the father cut in 1864 with the idea that he would use it when it came of age. It is the home of Gary and Tina Johnson on Chicago Road, south of Shabbona.

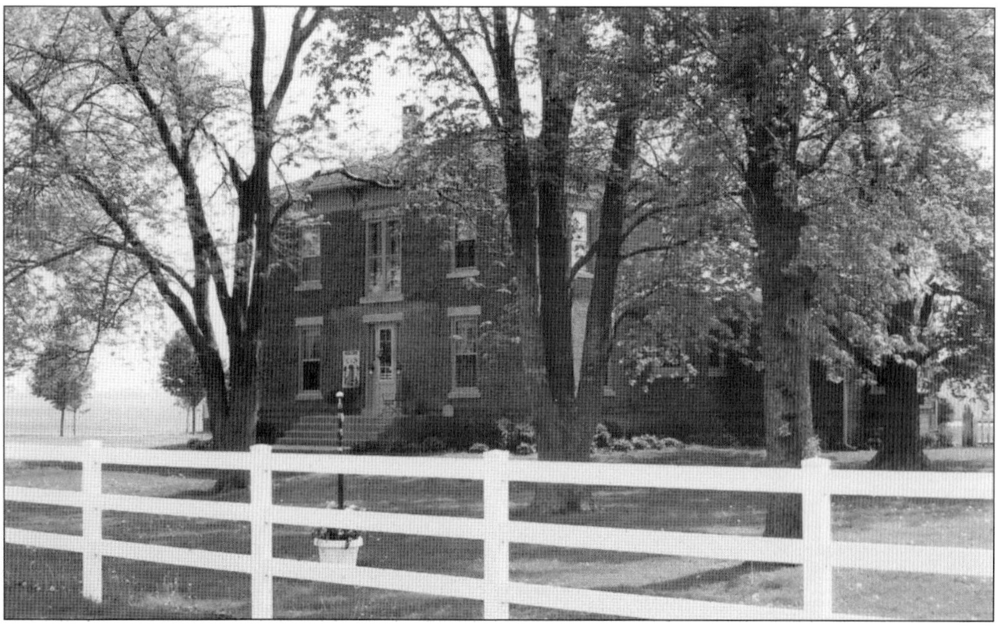

The former Virginia Hood home on Hyde Road, located southeast of Rollo, is a 12-room brick house that was built by Virginia (Ellsworth) Hood's great-grandfather Simeon Hyde in 1867. She grew up there and continued to live in the house with her husband and son. Blind since 17 years of age, Virginia enjoyed life and spent many hours knitting and listening to recorded books and music.

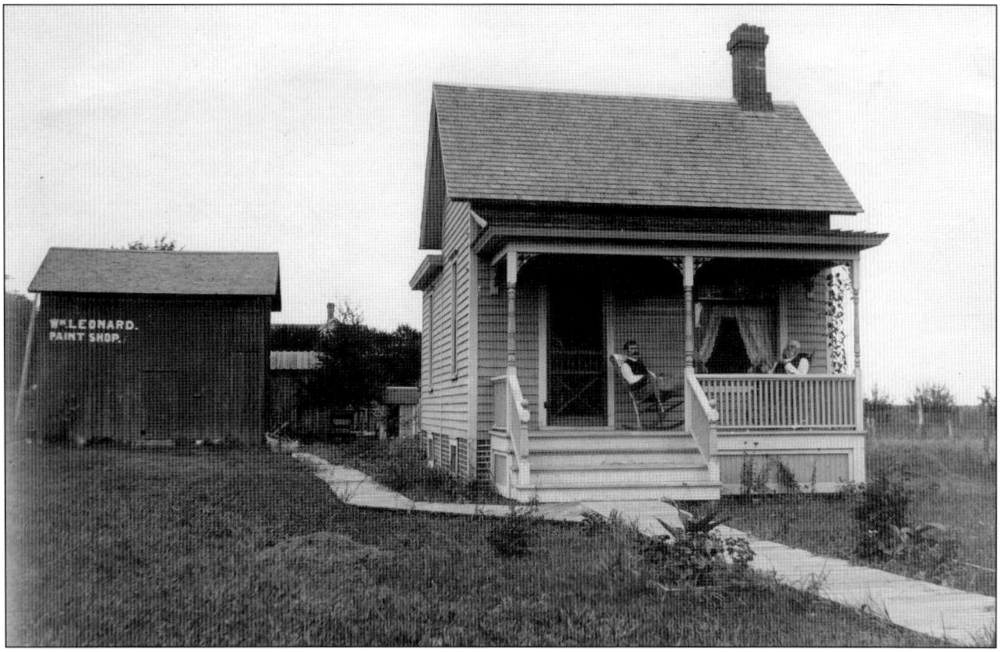

The original house and barn on the northeast corner of East Navaho and North Pontiac Streets belonged to William Leonard, who was a house painter. An addition was made to the house after 1908, which tripled it in size. It was owned by Percy Hallam and is now owned by the Ronald Larson family. Note the wooden sidewalk and the name on the barn.

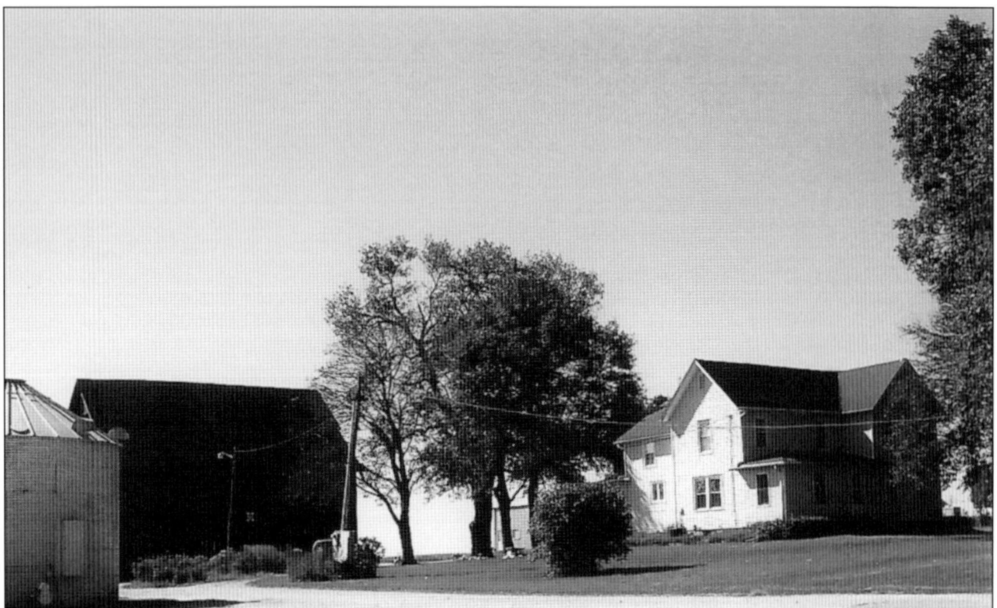

Just south of Shabbona Grove is a farm that has been owned by the Foiles family for over 150 years. The Foileses were innovative farmers, experimenting with many new practices of conserving the rich soil. Conservation tillage continues to be a high priority and means that the land will be suitable for generations to come. The home of Elsie and Eugene Drake is on Sleepy Hollow Road, south of Shabbona.

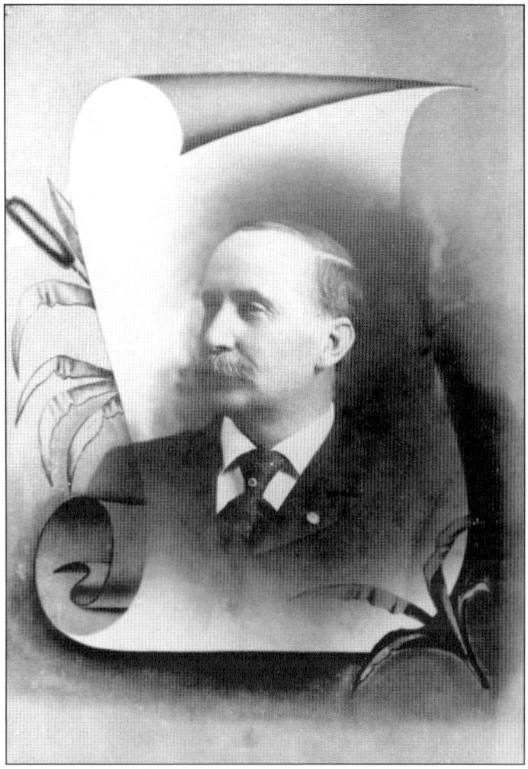

In 1877, a law was passed in Illinois, preventing anyone from practicing medicine until they had secured a license from the health department. The requisite for this license was that they should be a graduate of any approved medical school or should have been actively engaged in the practice of medicine for 10 years prior to the passage of this law. The early doctors, like the circuit riders, covered a large territory, and their monetary reward was small. Dr. H.A. York was an early physician in Shabbona Township. Dr. Robert Purdy, below, is a physician who came to Shabbona in 1953 and retired in 2005. He practiced out of his home for many years and also operated a clinic on Navaho Street in Shabbona.

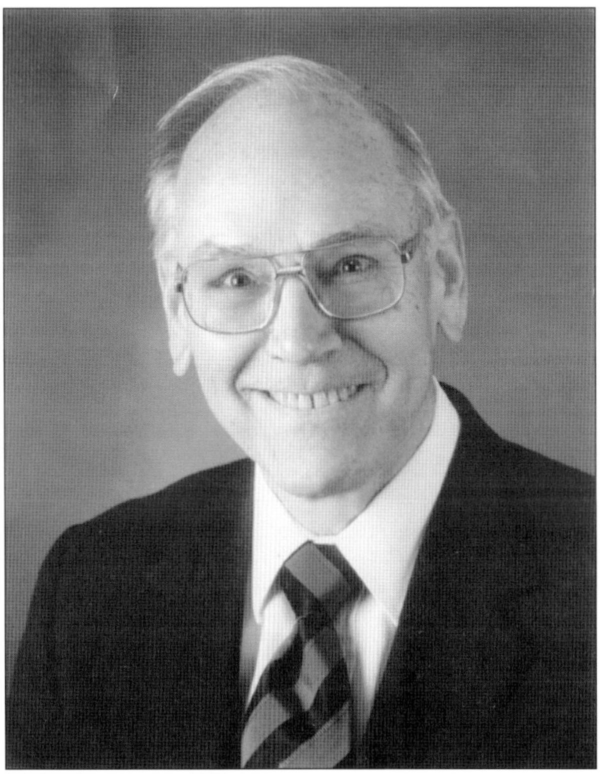

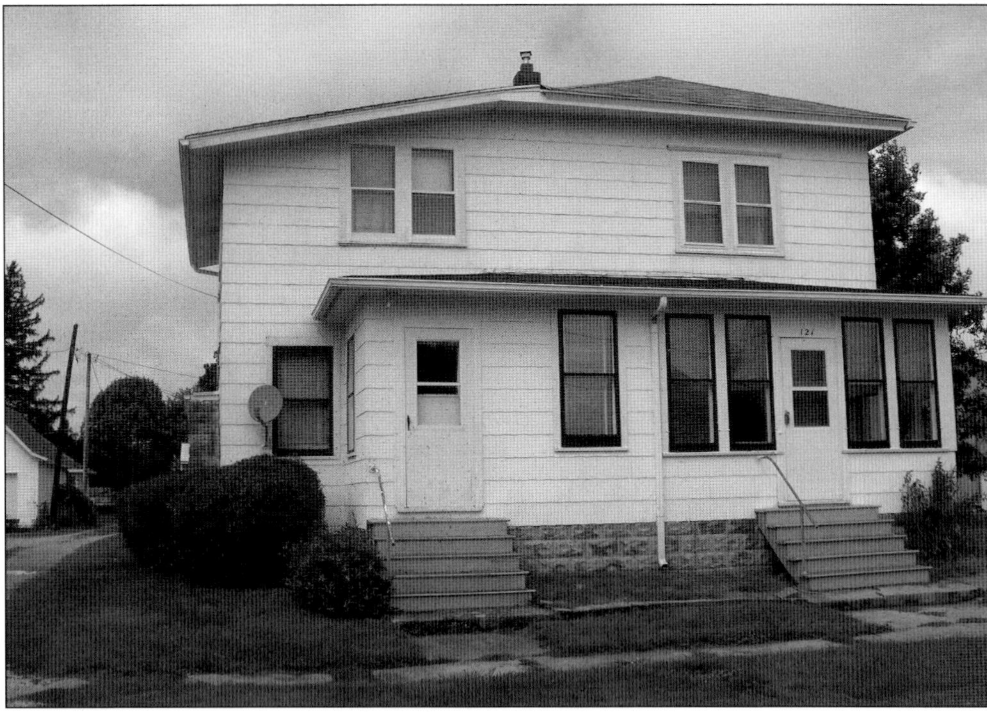

Dr. James Boardman was the first physician in Lee. Dr. Carl Axel came to Lee in 1933 and began his practice in the Maakestad Building, later moving to the location pictured above, which was where he and his family lived. He was called into military service in 1942 but returned to Lee in 1945 and practiced until 1961. He died of a heart attack while vacationing with his family and was the last doctor to serve the town of Lee. In the early years, Lee had two persons who practiced veterinary medicine: Hans Hayer was in Lee for over 25 years, having arrived around 1880, and Dr. Tysdale came in the early 1900s. Shown below, in a picture given to the museum by Helen Pickwell, are the offices of Dr. Hedberg and Dr. Edwards, who were early physicians in Lee.

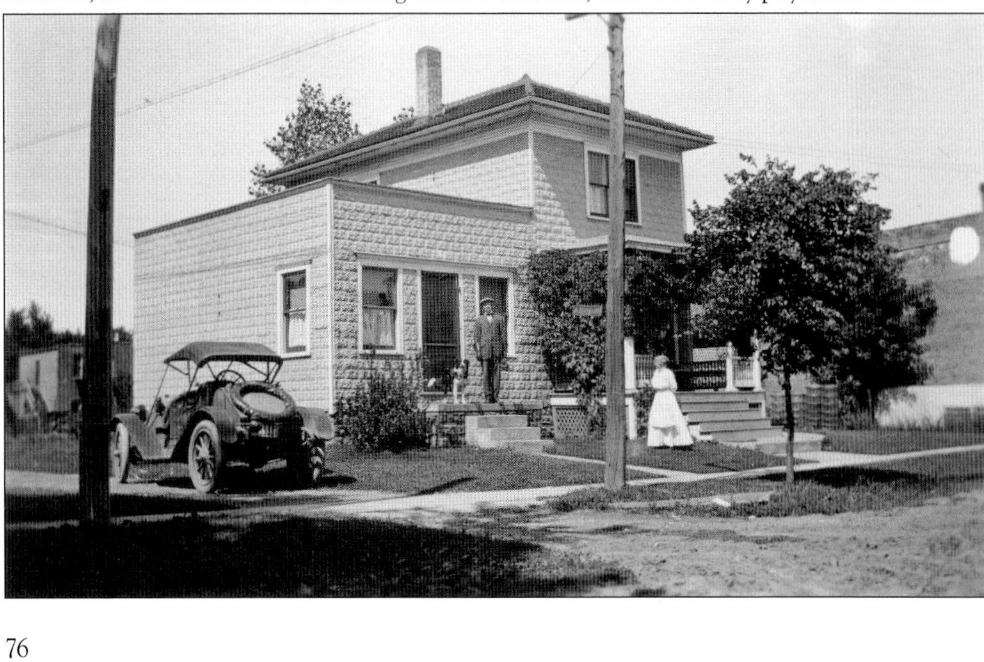

Seven

Main Street

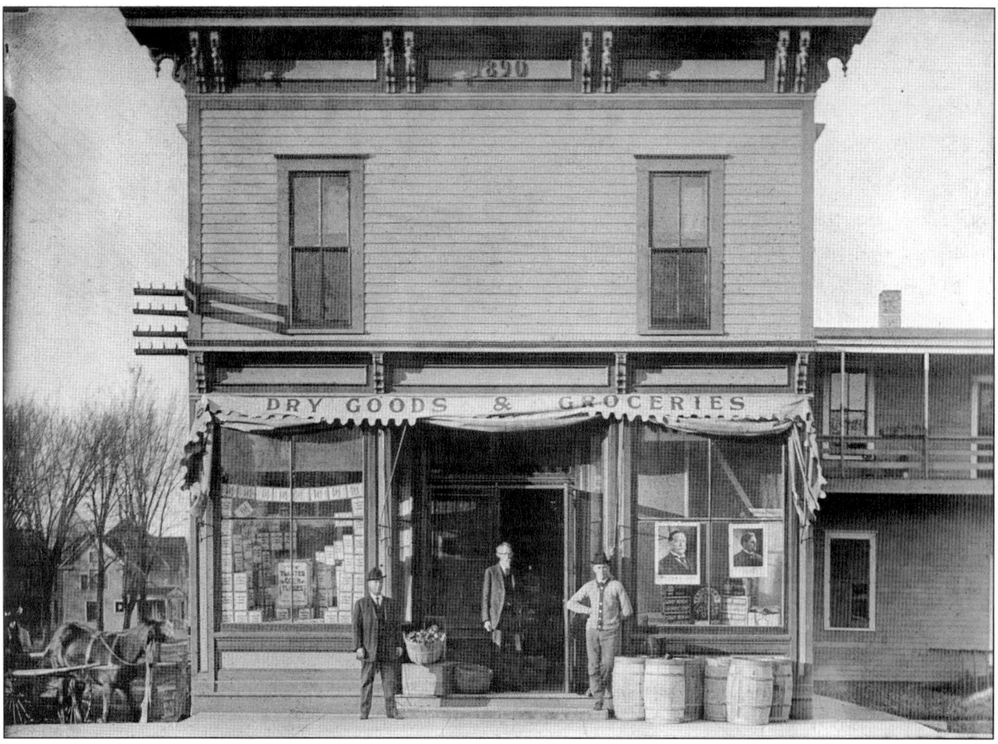

Over the years, numerous businesses in the small towns of the area provided solutions to the wants and needs of the people. With citizens working together for the good of the community, the businesses moved from the small towns to where the railroad would be located. In Shabbona around 1905, Wells and Larson ran a general merchandise store. Note the sign in the window for William H. Taft.

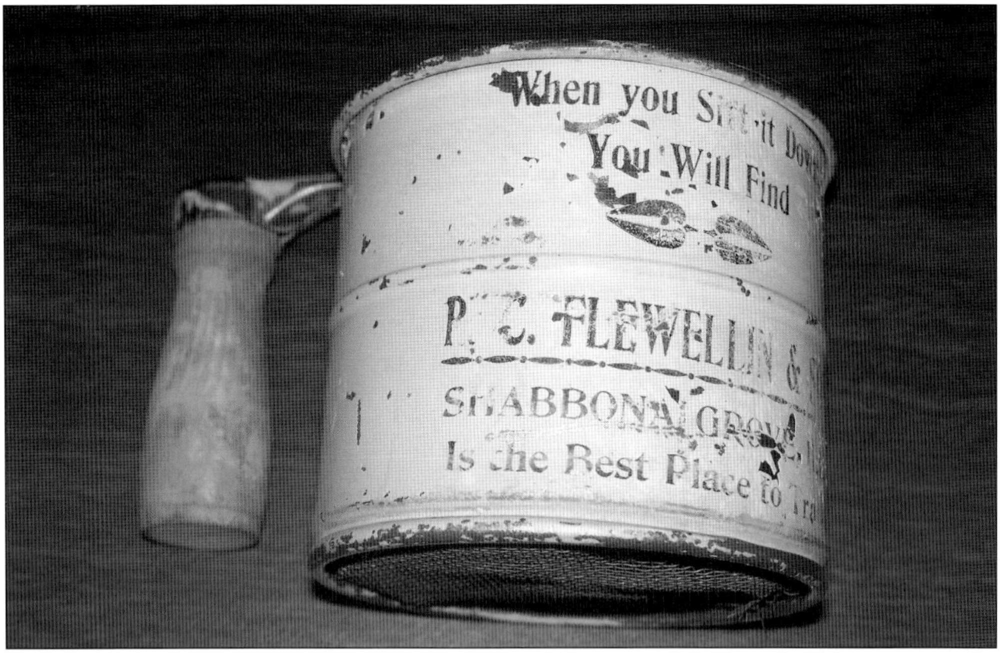

In Shabbona Grove around 1915, R.C. Flewellin and son ran a general store and operated a post office. The two-cup flour sifter (donated by Jerry Maple) is from the store and has the wording: "When you sift it down you will find R.C. Flewellin and Son, Shabbona Grove, Illinois, is the best place to trade." It is on display at the Shabbona-Lee-Rollo Historical Museum.

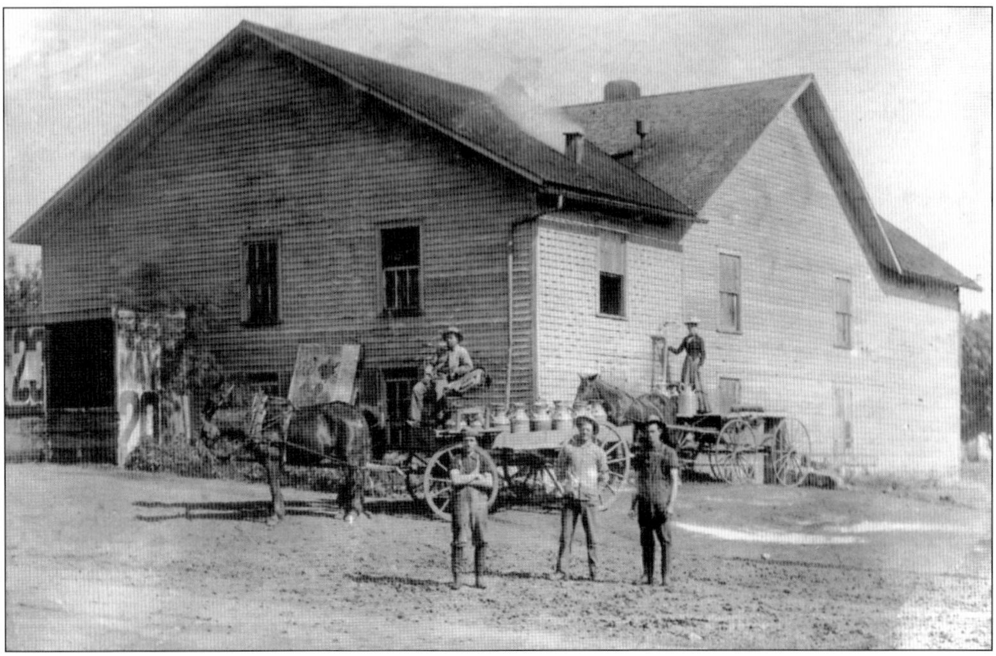

The old creamery was located in the eastern part of Shabbona on the north side of the main street where Oscar Olson's house stood in 1951. The creamery was in business for quite some time, containing the latest machinery in the line of separators, churns, and so on for that time period, around 1932. A Mr. Gurler ran it. The picture was given to the museum by Doreen Smith.

Dr. Edwin Brewer came to Shabbona to practice medicine in 1901. He was born in Ashton, Illinois, and attended the University of Illinois. He had been a professional boxer and kept a picture of himself in his office. When the moving pictures came in the old silent days, Dr. Brewer bought the opera house in 1905 and showed movies in Shabbona for years in the days when movies brought crowds to the small towns every night. Brewer's Opera House in Shabbona provided entertainment of all kinds until 1933, when fire broke out, and the building was destroyed. Dr. Brewer was interested in his community and took an active part in its affairs. He was mayor of the village of Shabbona from 1923 to 1927 and from 1929 to 1935. He died in 1939 and is buried in Ashton, Illinois. the picture at right was given to the museum by Helen Pickwell.

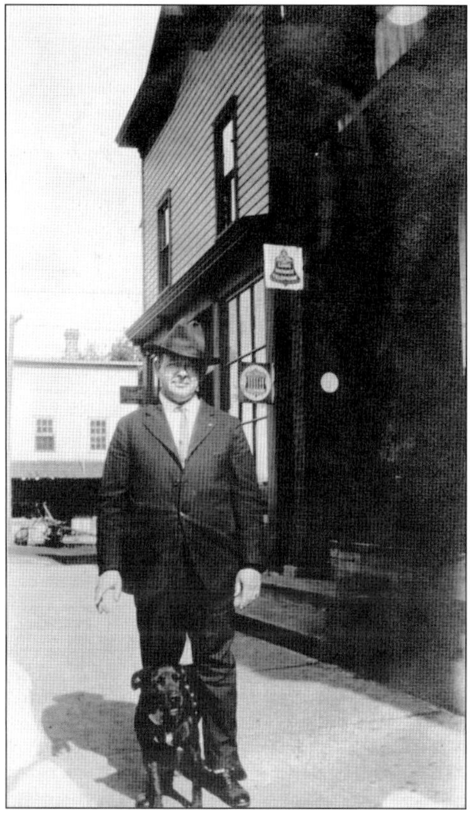

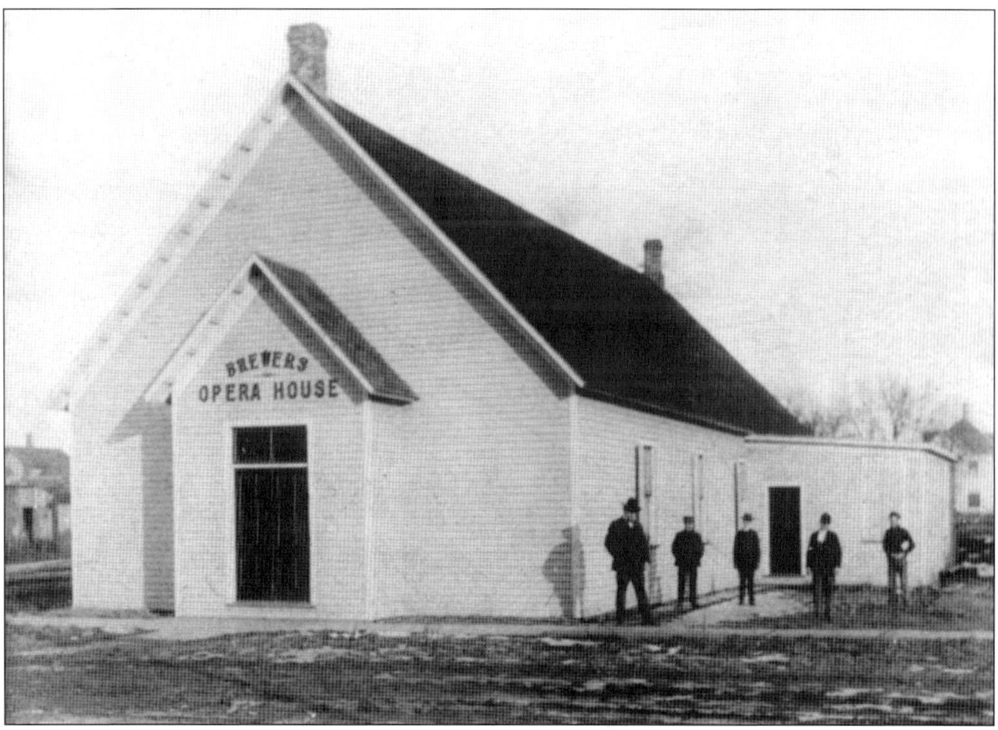

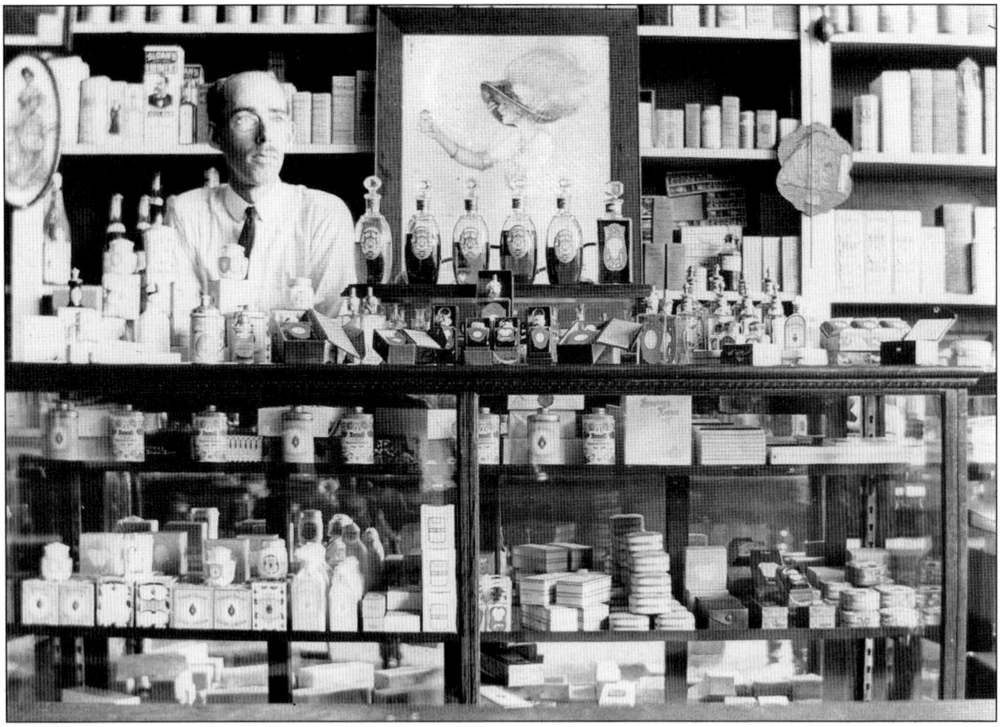

C.P. Hallam operated the Rexall Drugstore at the southwest corner of Illini and Comanche Streets in Shabbona. Capt. M.V. Allen opened the original store in the fall of 1872. The family had purchased the building from Dr. York in 1907. In 1923, the Svihus brothers erected a modern brick building on the same location for Percy Hallam. The bricks came from the old school, which burned. Percy hired young boys, including Millard Thorsen, to clean the bricks for 2¢ a brick. The building is well designed, with cement for the floors, making it somewhat fireproof. It had a drugstore, including a soda fountain, on the main floor and two apartments upstairs. He served the community from 1919 until he retired after 51 years. Emma Hallam retired from teaching after 20 years. (Both, courtesy of the Hallam Family Collection.)

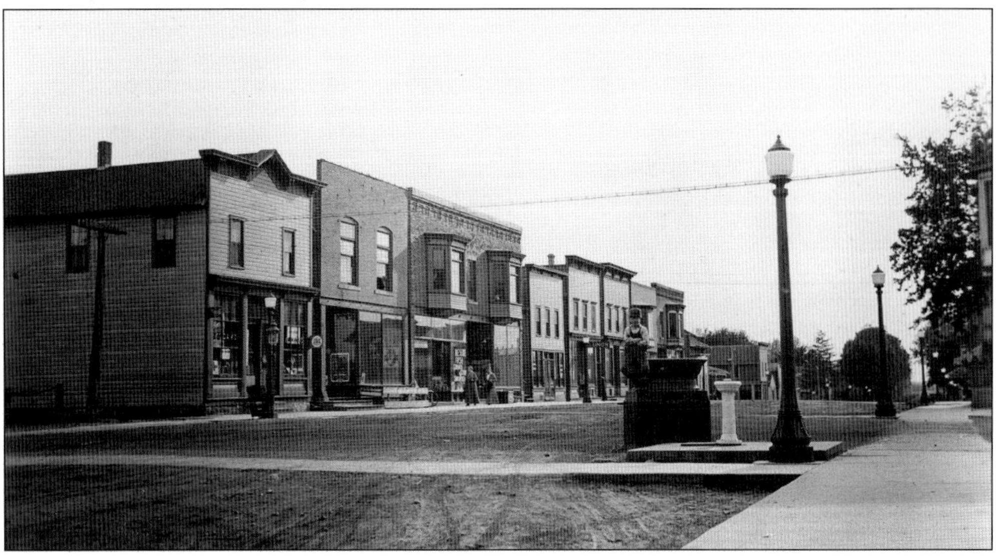

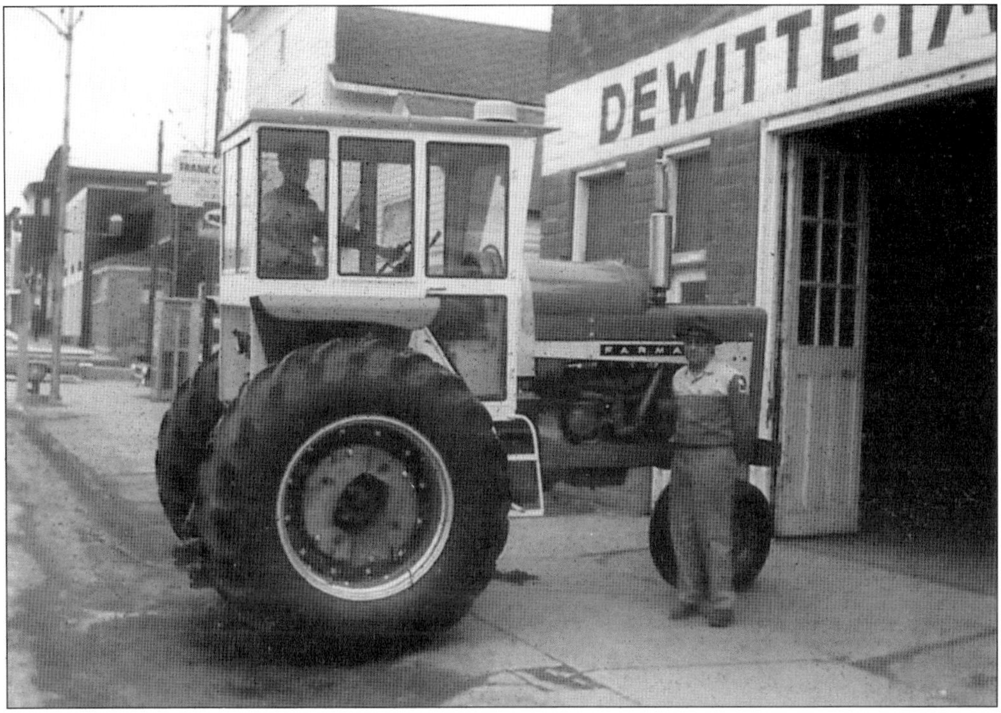

DeWitte Implement Company, run by Maurice DeWitte and family on Main Street in Shabbona, purchased the business of the Swanson Implement Company on September 15, 1950. An open house for all local farm folks was held on Thursday, February 3, 1955, at the DeWitte Implement Company to introduce the first of the line of five new Farmall tractors for sale. The new tractors constituting 16 models were announced by International Harvester Company in what the company said was its biggest tractor development since the first Farmalls were produced in 1923. New features included the Farmall Fast-hitch, more powerful engines, new or improved hydraulic systems, independent power takeoff, and new operator conveniences for the driver.

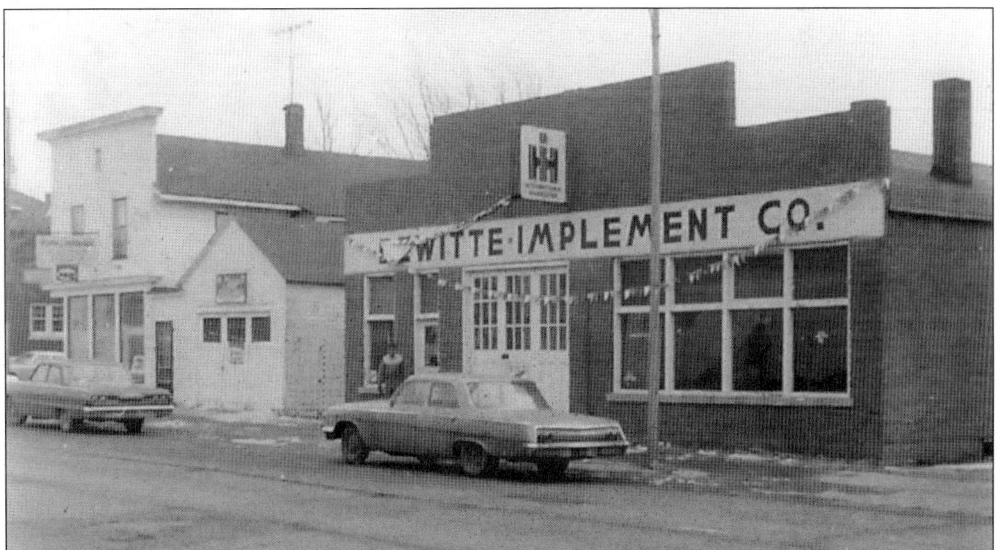

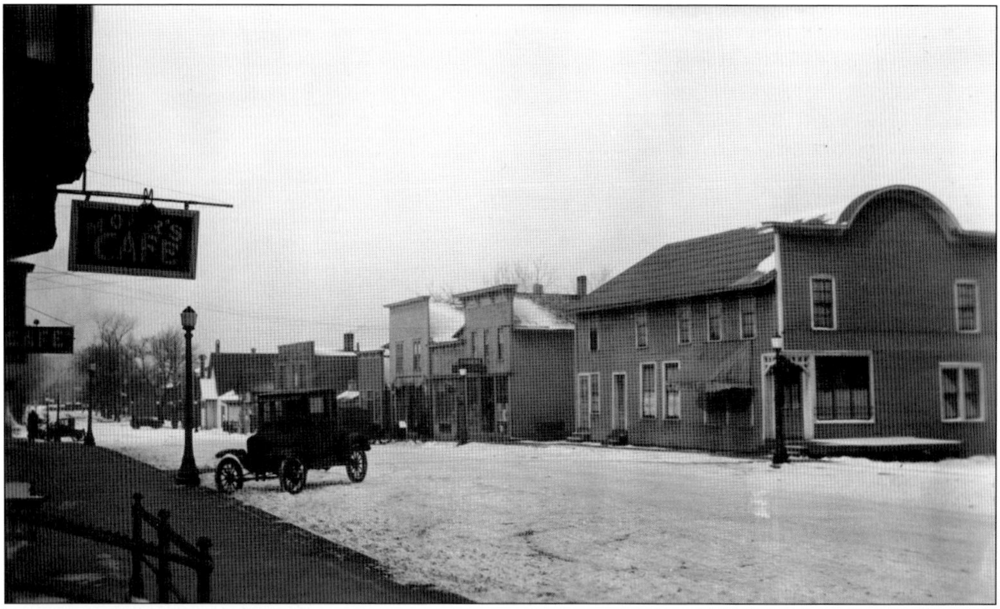

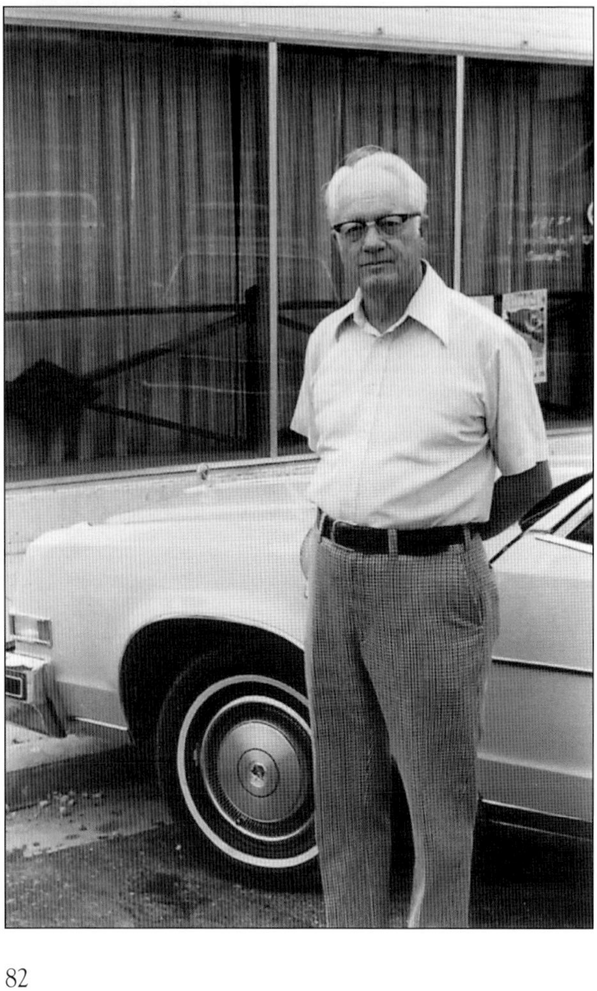

Howard Wirt and family moved to Shabbona in 1915 from Nebraska; his father was born on a farm near Paw Paw, and his parents were members of an Illinois pioneer family. He operated Wirt Funeral Home and Wirt Furniture, purchasing the store from William Heeg. When the Wirts left Shabbona in 1927, the store was sold to Adam Heeg, who in turn sold it to Millard Thorsen in July 1940.

Millard Thorsen's father, Martin, was born in Norway. He left Norway to seek out farmland in the United States. He married Martha Olson in 1895, and Millard was born in Shabbona on the Henry Husk farm southeast of town. Thorsen erected a new building in 1955 for the furniture store and made changes to his home in order to convert it into a funeral home.

Shabbona Express was purchased by C.O. Trytten in 1928. It was located in one of Shabbona's older buildings at the northwest corner of Illini and Comanche Streets. The building has had many uses over the years, including as a schoolroom and dry goods store. Doreen Trytten Smith bought the building in 1957, and the paper, which was renamed the *DeKalb Co. Express* in the 1920s, was discontinued in 1960.

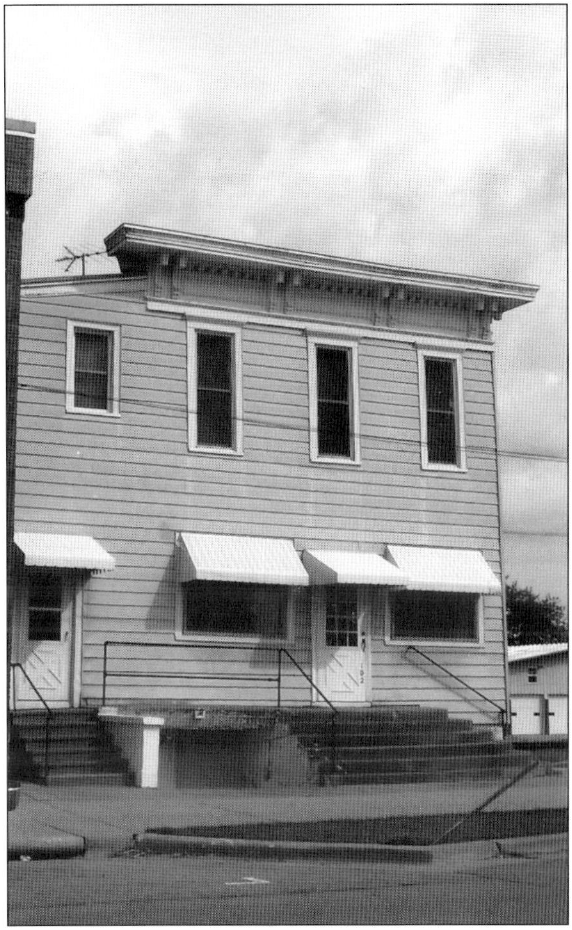

The Shabbona Hotel, below, was purchased by Russell and Judy Hogan around 1965. C.W. Hubbell erected the building, which was the second building from the corner of Comanche and Illini Streets in Shabbona in 1899. In 1938, the building was sold to Les Derr, who operated the barbershop and pool hall until 1953, when LeRoy Kitner purchased it. The building has been a barbershop, hotel, apartment building, and boardinghouse.

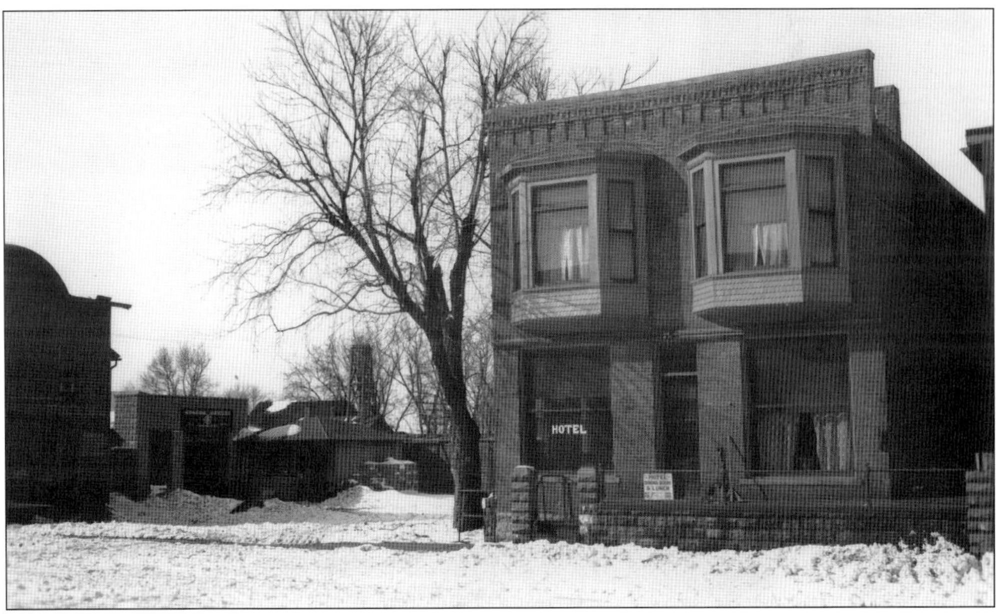

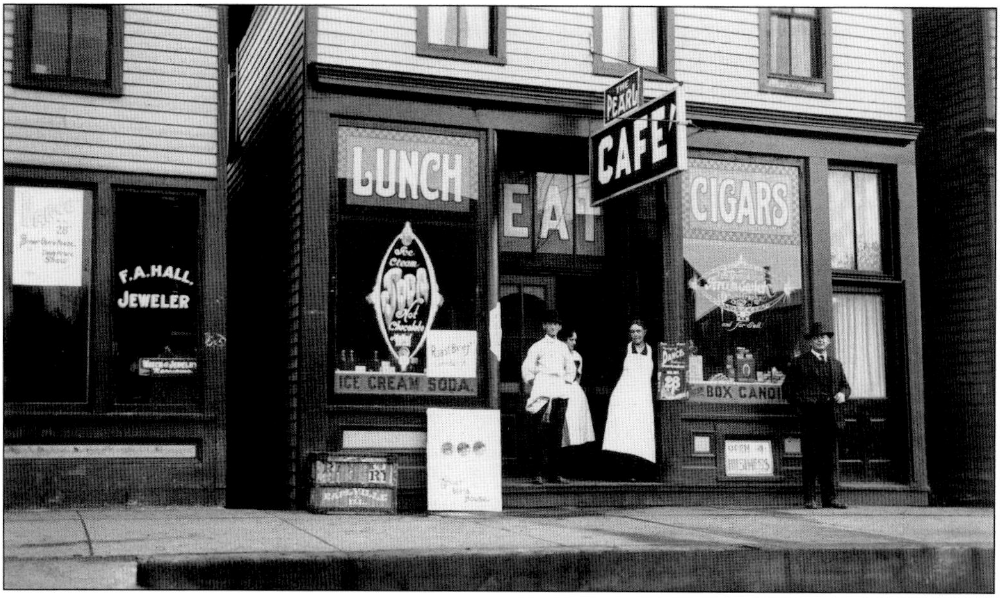

In 1910, the fifth building from the drugstore at the corner of Illini and Comanche Streets on the south side of the street in Shabbona was the Pearl Café. F.A. Hall Jeweler was next door. In the 1950s, this building became the restaurant of John Brockhaus. John came to town from Wisconsin to work for the Umland Cheese Factory. From 1943 to 1957, he and his wife, Doris, operated the Brockhaus Restaurant. When the restaurant closed, John went to work at General Electric, and Doris worked for Dr. Robert Purdy for almost 30 years. (Both, courtesy of Farmers and Traders State Bank.)

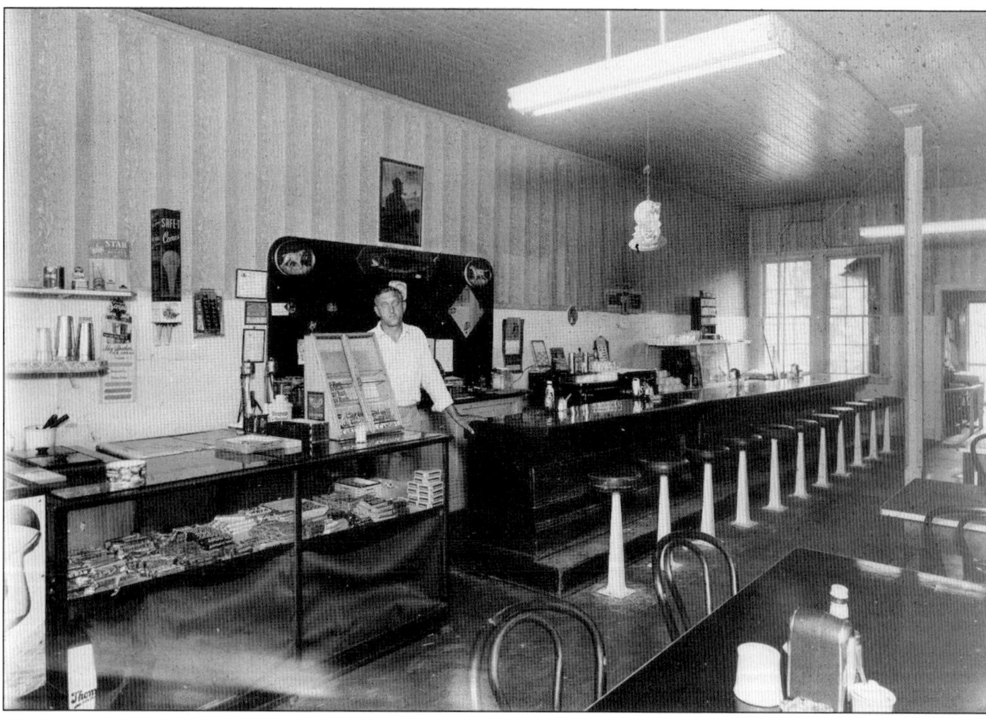

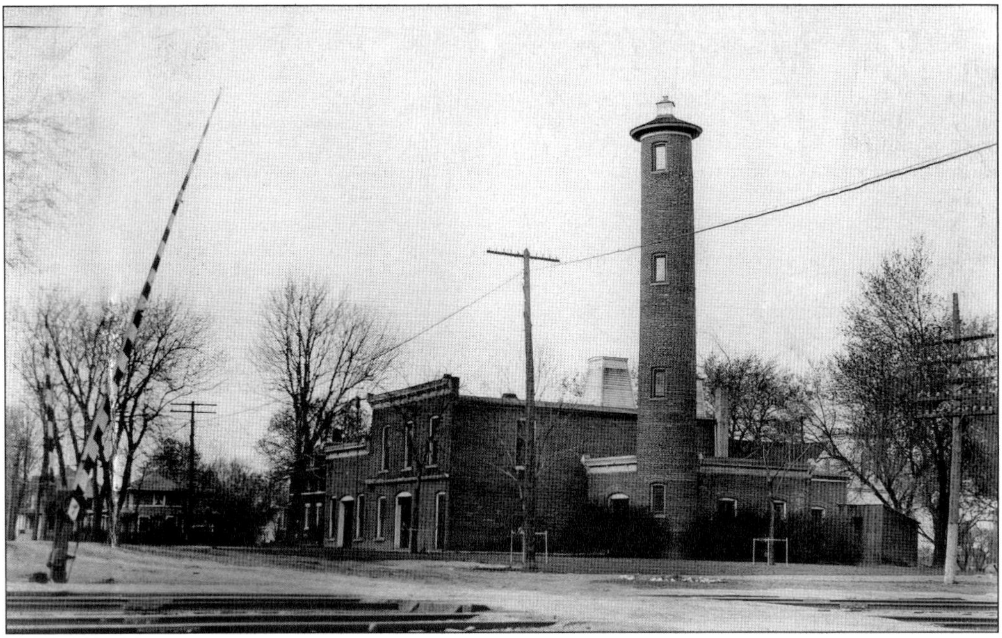

In a guest column written for the *DeKalb Chronicle* in 1993, Fritz Vandre recounted how the pump house was built and served Shabbona for many years. In 1900, the water mains were put into the new building on the southeast corner of what are now Navaho and Illini Streets. The tower was over 40 feet high and contained two wells and pumps. In 1940, the tower was reduced to 25 feet, and in 1986, volunteers repaired the building. The waterworks building was home to many organizations in its 100 years of existence, including the jail. The new water tower was completed in 1969, and all that remains on the site of the original landmark building is the Veterans' Memorial. Across the street, the depot has been restored, and the repaired water tower looks over the village.

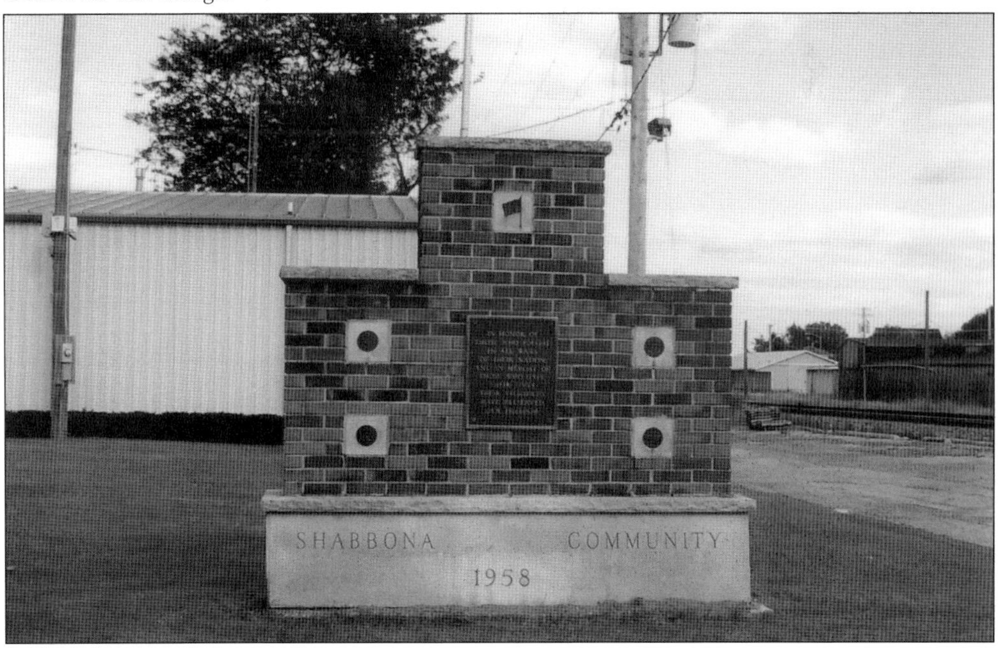

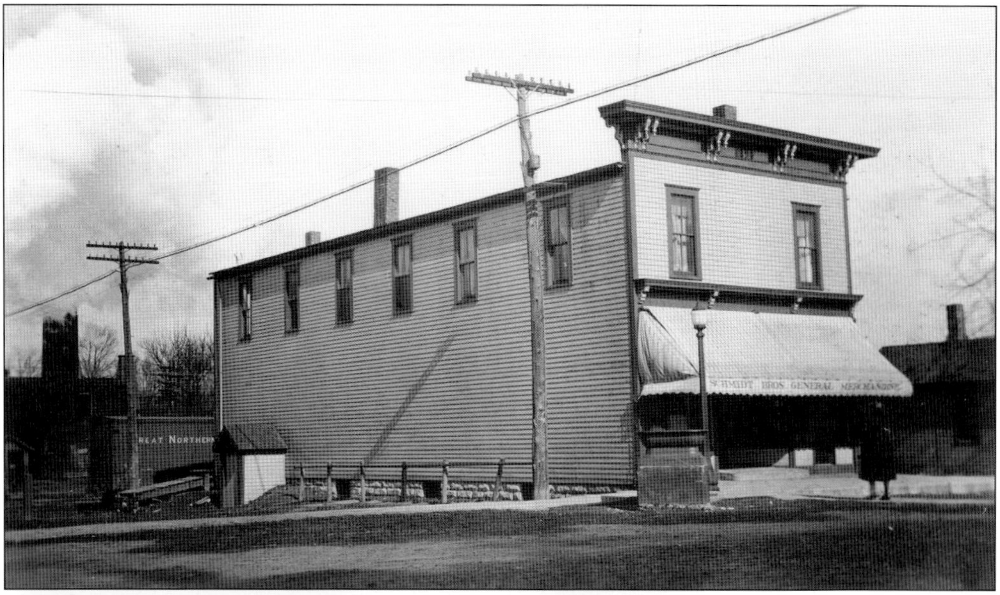

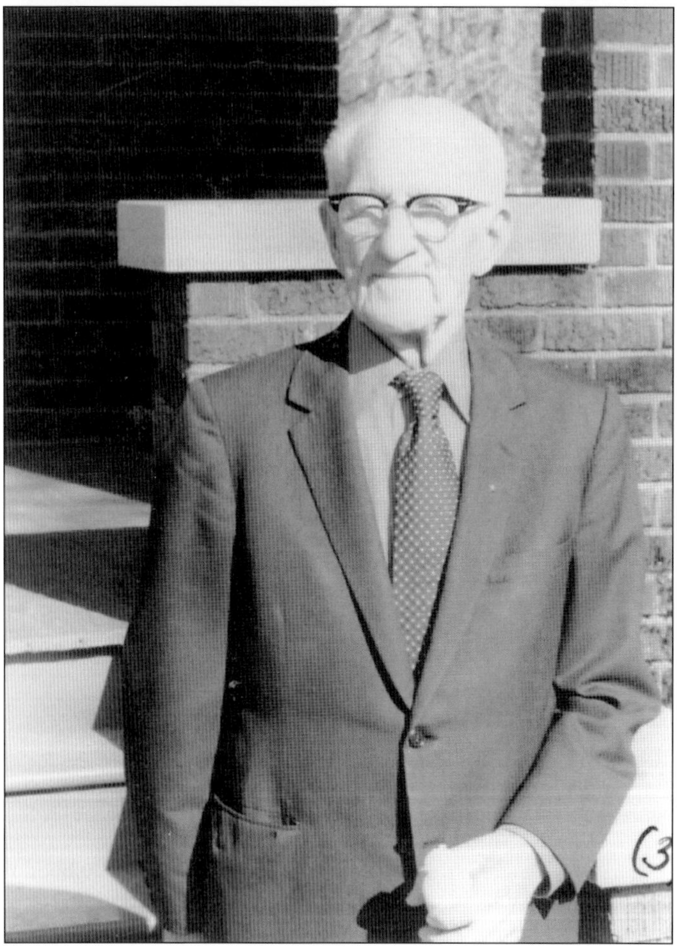

Otto Schmidt and his brother Ray came from Hinckley, Illinois, in 1912 and opened a dry goods and grocery store in the Swanson Building on the corner of North Illini and Comanche Streets in Shabbona. A few years later, he bought out his brother. In 1920, Schmidt bought out Chris Munson's stock upon his retirement and moved to the south side of the street. George Jackson and Captain Lucas constructed the building in 1898. It originally housed a hardware business. He discontinued his line of men's clothing when Lars Solberg started a men's clothing store in Shabbona. He sold his stock to Clare O'Kane in 1953 and retired. Note the gaslight on the corner and the watering trough for horses.

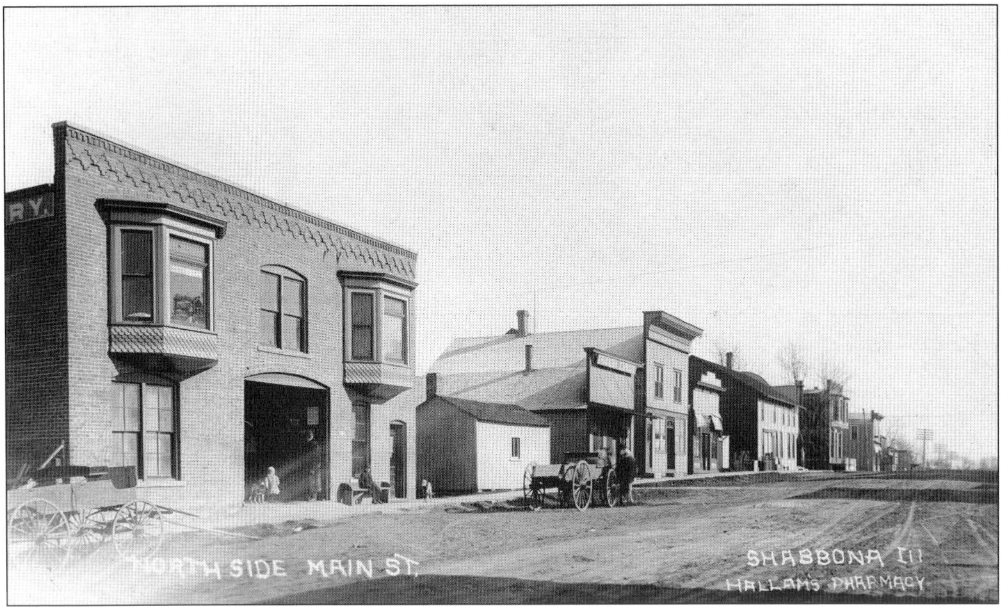

This picture looks east on the north side of Main Street in Shabbona. The first building was George Challand's Garage, which is now a gun shop. The second building is now a restaurant, formerly the Legion Hall. The third building was the *Shabbona Express*, and the fourth building was a millinery store. The fifth building has a sign above that reads, "Billiards and Pool."

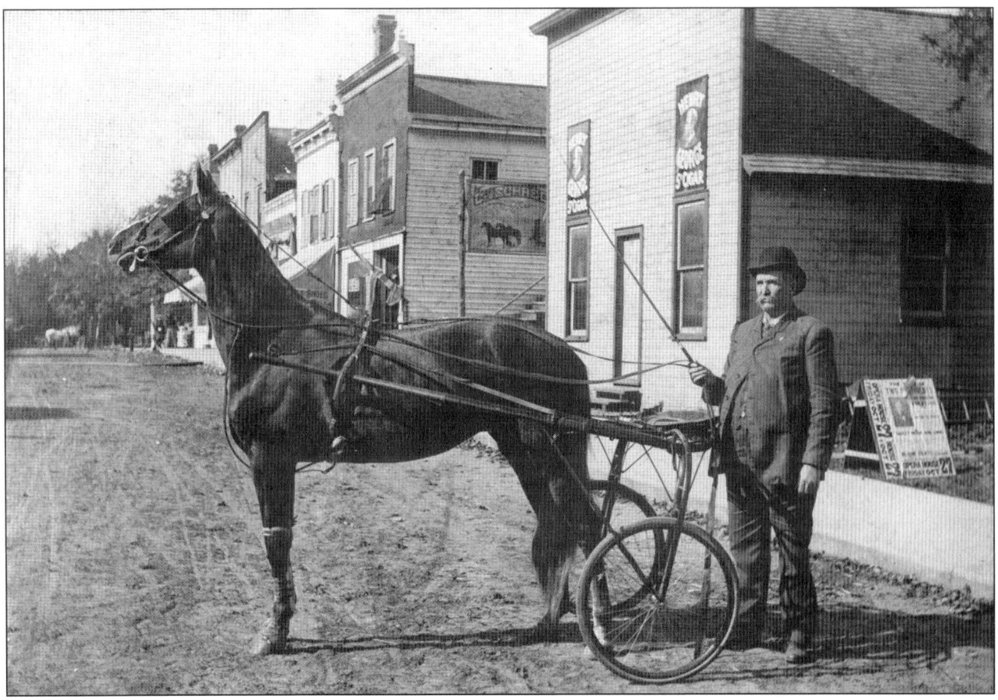

Wellington Helm and his pacer were regular winners at the racing park south of Shabbona. His parents came to Shabbona in 1852 and purchased a farm near what is now the village of Shabbona. He was married to Sarah Van Velzor, and they lived on the farm for a few years before building a house in town.

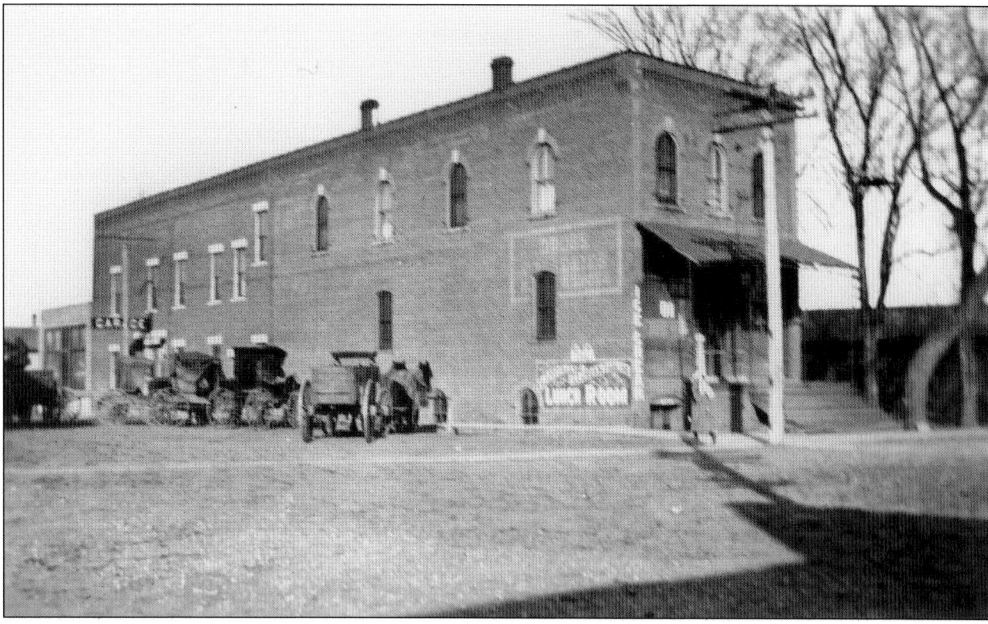

In 1881, Sardinas Vosburgh constructed the first brick building in Lee. It was used for Nelson's Opera House (Woodman Hall), roller-skating, plays, silent movies, and other forms of recreation. Earl Wrigley and his son Gene had garage businesses in the building. The American Legion Post bought the building in 1955. The building was sold in the late 1980s to Cass Larson and is now an apartment building.

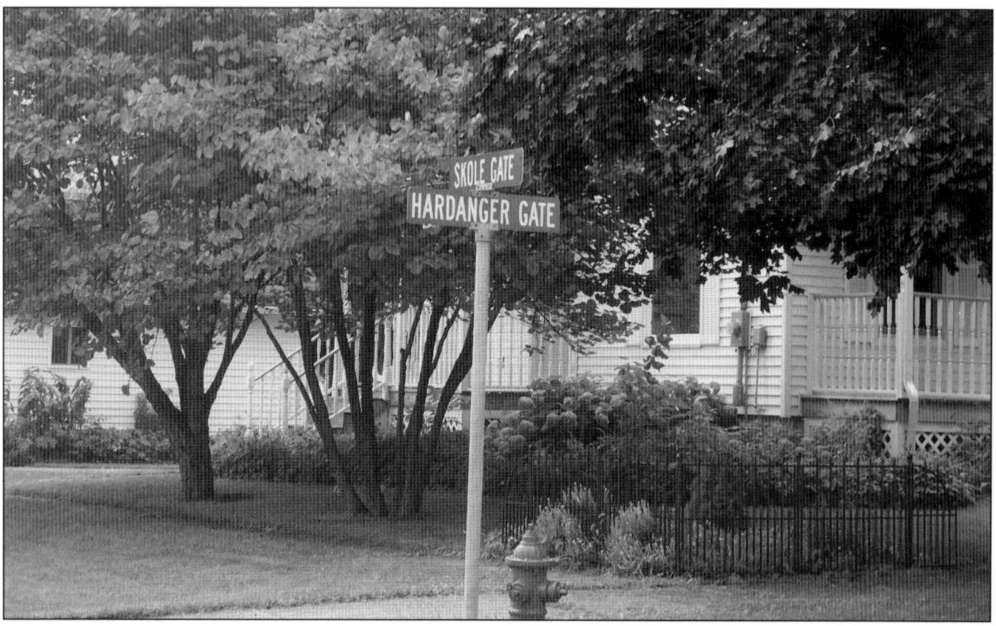

After 99 years of rural route addresses, Lee named its streets. In anticipation of a centennial celebration in 1974, a contest was conducted, and the winner was Mary Ellen Prestegaard of Lee. All of the streets are named in Norwegian and according to each one's features. The photograph is of the corner of Skole Gate (School Street) and Hardanger Gate, a district in Norway from which many of the early Lee settlers had come.

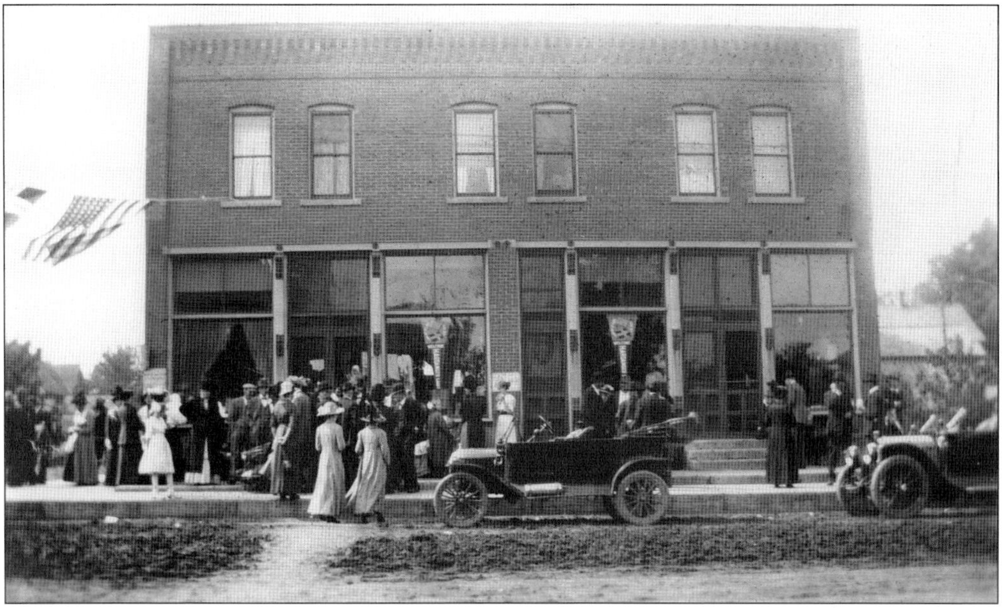

Art and Doris Erickson operated Royal Blue Grocery Store in this 1903 Maakestad Building. Since coming to Lee in 1946, they also purchased property for a packing plant, beauty shop, restaurant, and a barbershop building, which was remodeled into several businesses. Erickson served as the mayor, president of the county board, and township supervisor.

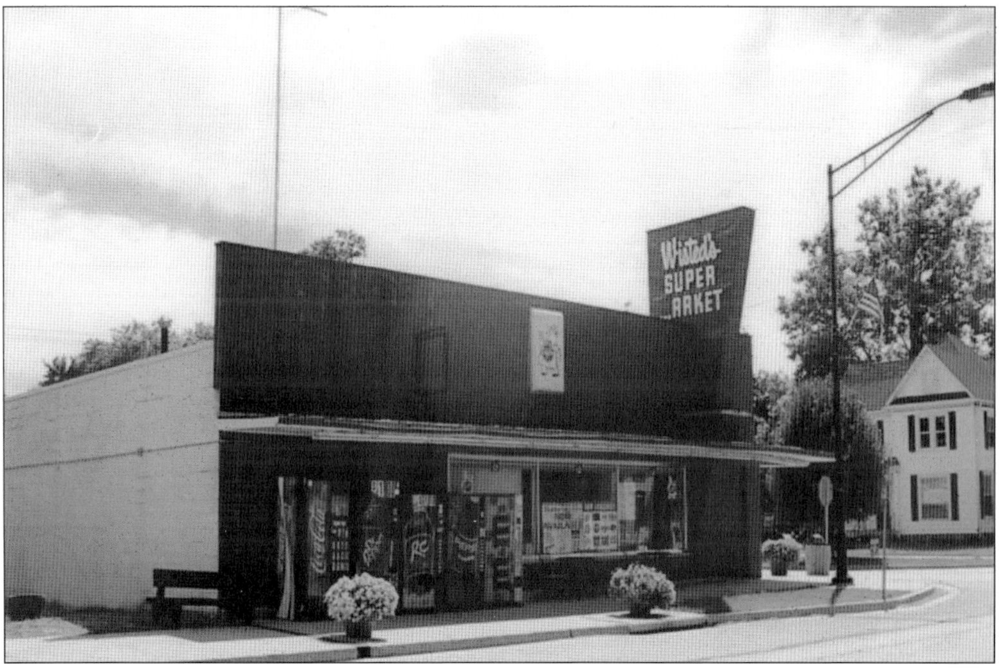

Wisted's Super Market, located at 102 South Nakomis Street, was the former site of a building erected by Mr. Hutchins for a blacksmith shop. David Jones purchased the building from Hutchins and operated a blacksmith shop. Dr. Harry Smith bought the building and later sold to William Jacobson. In 1960, Charles Wisted bought the building, had it torn down, and erected the present building for the family super market.

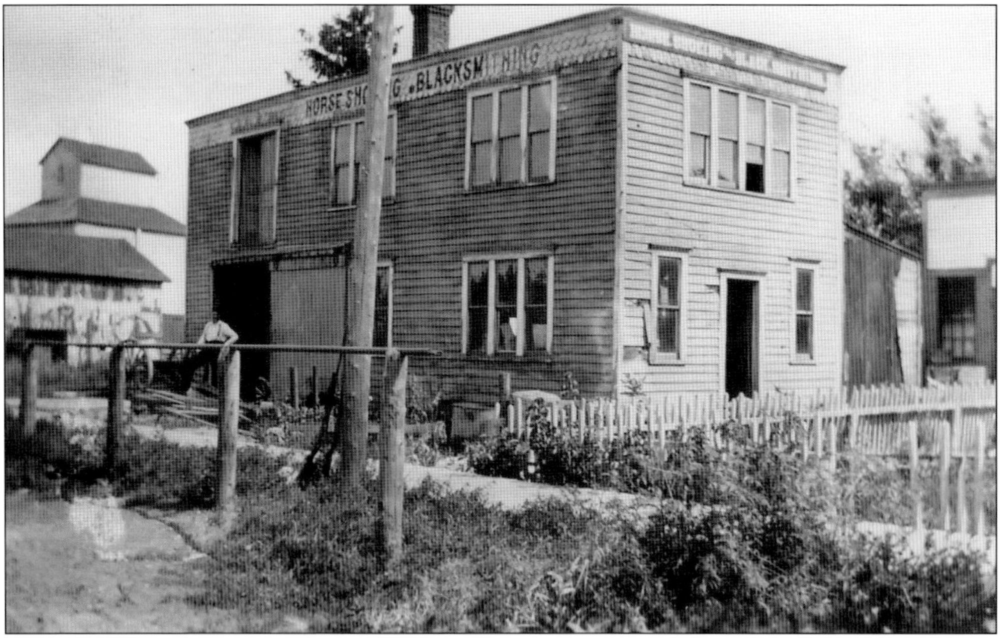

Three blacksmith shops were thriving in Lee in the mid-1900s. They were owned and run by Lars Jordal, John Malmberg, and John Tonch. The Malmberg shop continued in business until 1943, when he sold to C.N. Rynearson and Son. Alvin Jacobson built his shop in 1954 and sold out to Phil Foster in 1963, which became Farmers Factory. It is now closed.

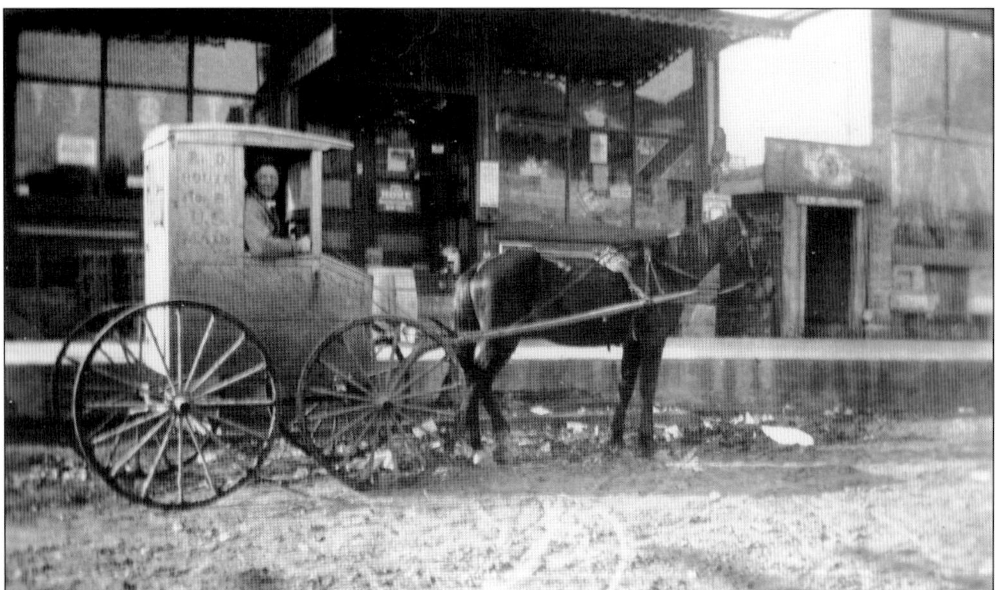

This early photograph of Lee shows a rural postal carrier ready to deliver the mail with his horse and buggy in front of a grocery store. The carrier would use his own horse and buggy and build a box on top of the cart to deliver the mail. Rural Free Delivery (RFD) began in the United States in 1896. The wording on the side reads, "R.D. Route, US Mail."

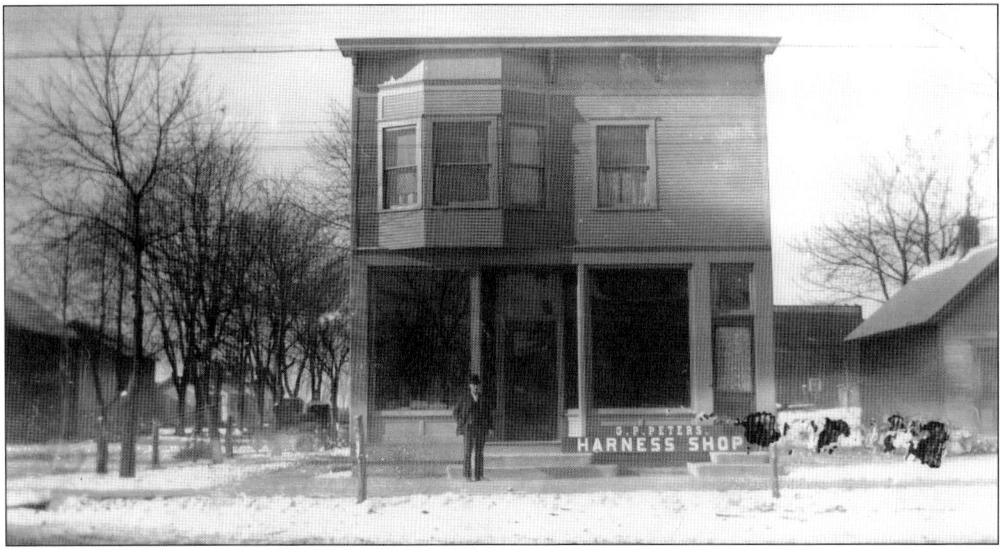

Every early village had a harness and buggy shop. Horses were the only means of transportation for the early settlers. The harness maker was responsible for making all the saddles and harnesses residents would need to hitch up a team of horses. The G.P. Harness Shop was located in Lee before 1908. John Whitby followed Mr. Peters in 1924 at the same location. The site eventually became Gochee's TV and Radio Repair.

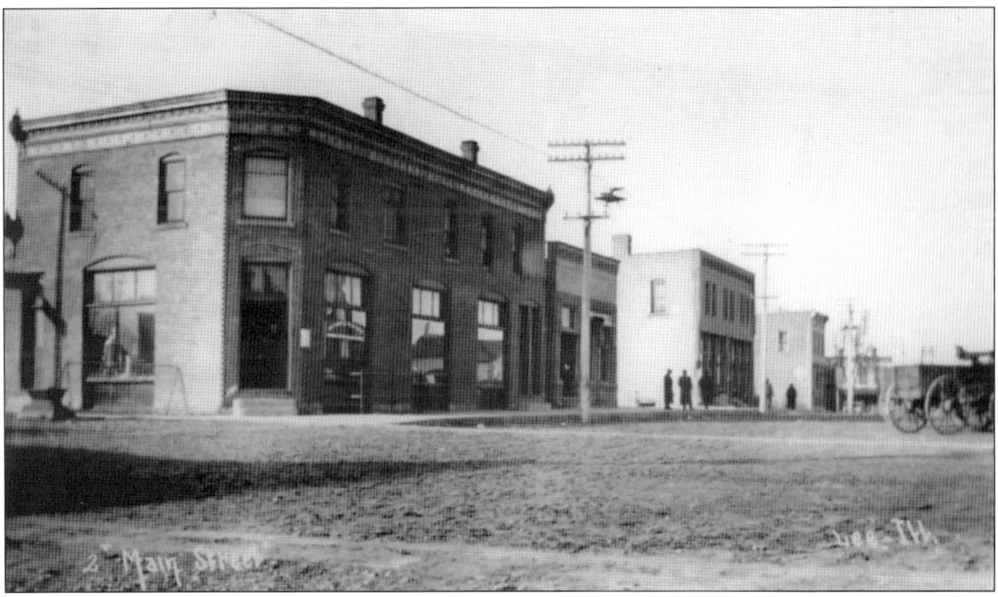

In July 1902, the wooden store buildings on the main street in Lee were swept away by fire. D.C. Henderson built this corner brick building where Merle Pine had numerous businesses, including his rock and mineral shop and his weed and spraying business. Next door was the bank, which closed in 1931.

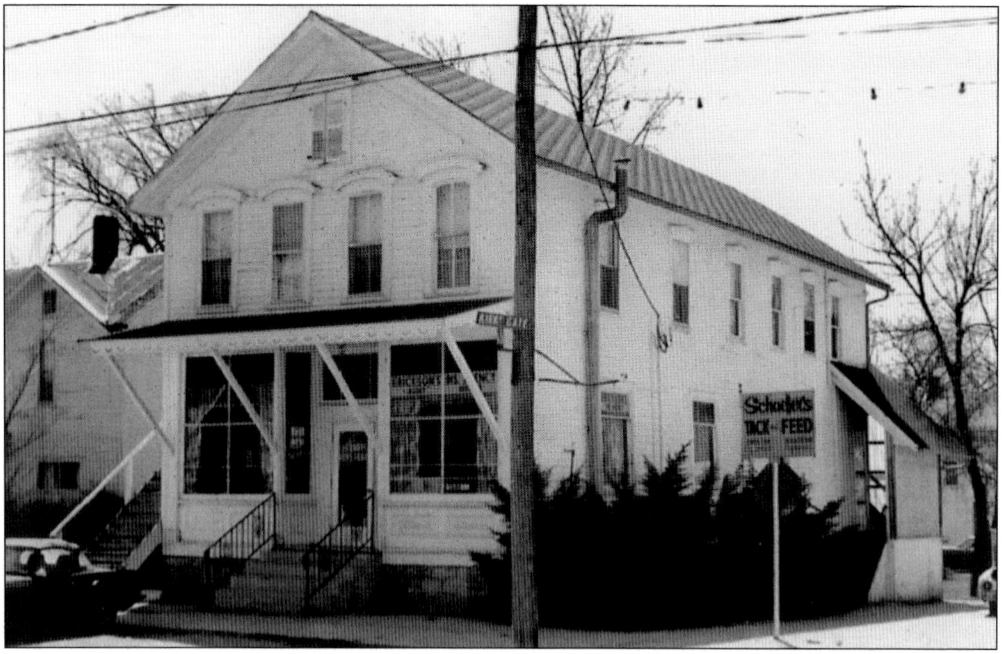

There were several tenants in the Risetter Building over the years. He also had a restaurant. Guy Irving came to Lee in October 1933. He bought the Risetter Building and operated his barbershop for 30 years until his death in 1963. The Eide Hardware Store and Erickson's Insurance Agency were also in this building for many years. It is now an apartment building.

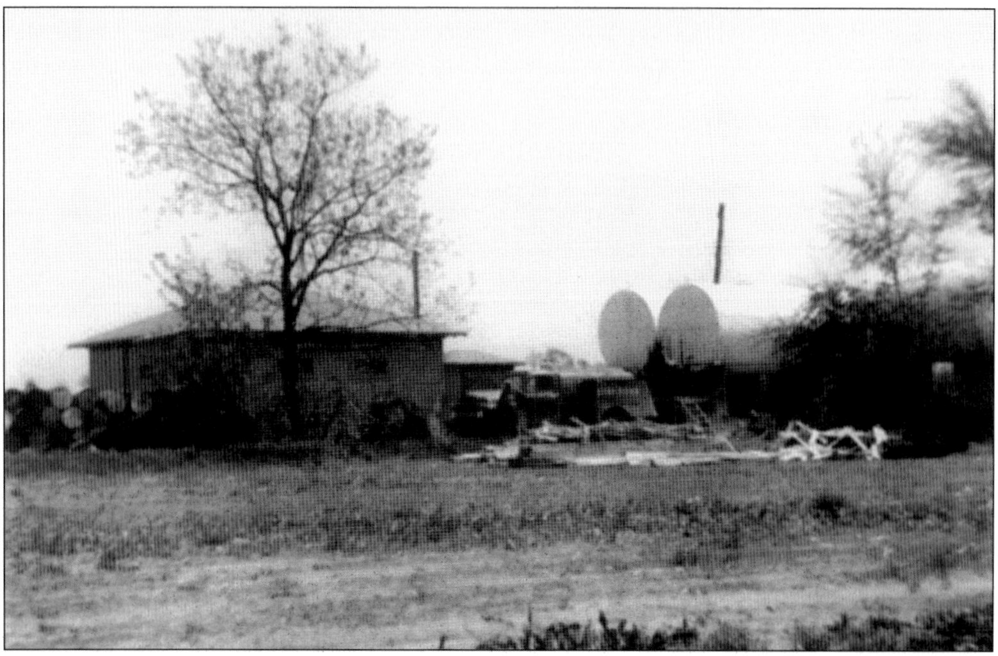

The Standard Oil Company maintained several bulk plants and numerous terminals and stations around the state. Standard Oil Bulk Plant in Scarboro, Illinois, was constructed in 1919. Ellis Rees helped in the construction and delivered from the plant for 42 years, at which time it closed. (Courtesy of Shirley Rees.)

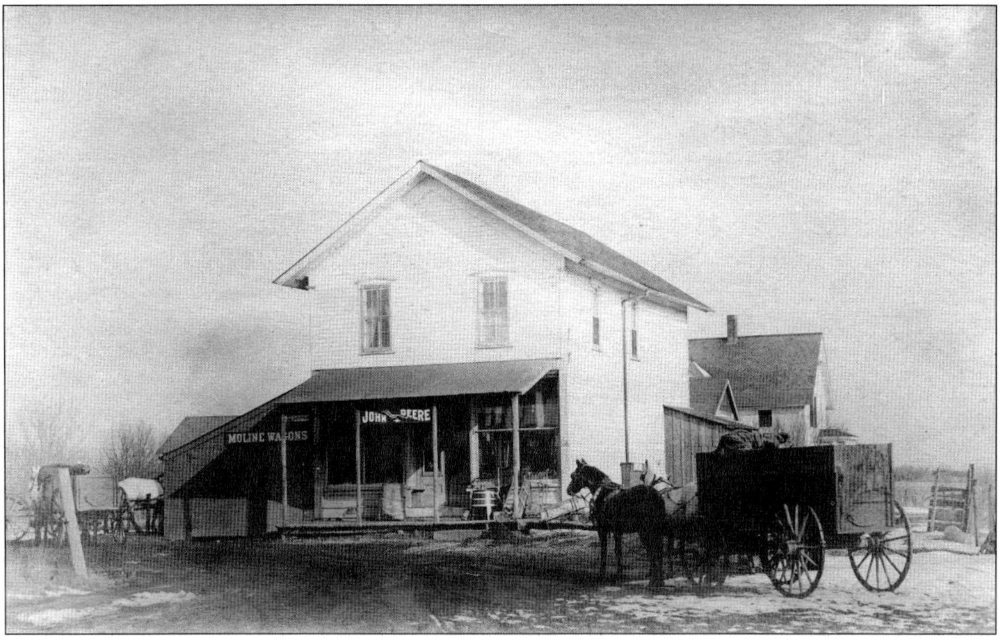

The date of the first store in Rollo, Illinois, is uncertain, but in 1888, Dupee Brothers and Hicks had a store, which sold dry goods, boots, hats, caps, crockery, and hardware. There was a small glass case of candy, and on days when the storekeepers had a tub of ice cream shipped in, there were ice-cream cones. Fire destroyed the old store, and Harvey Bullis erected the brick building, shown below around 1912, which still stands today. The new store had a long marble-topped counter. Customers could buy anything there, from mousetraps to calf weaners. From miles around, all the men in the neighborhood could come after chores and discuss politics and the war. There were cracker boxes and barrels enough for all. The Rollo Co-op closed on January 27, 1973, after over 60 years in existence.

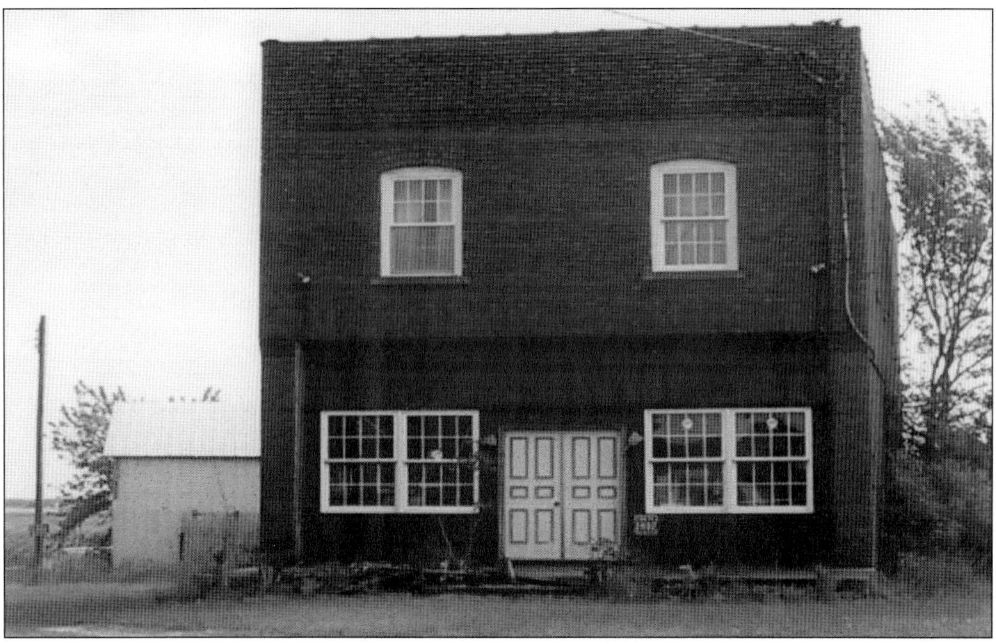

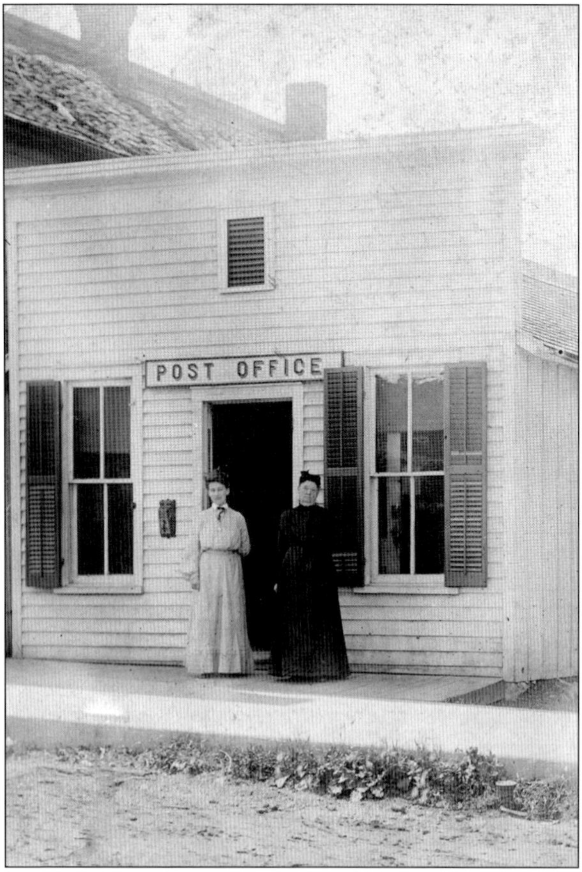

The first post office in Shabbona was in front of the general store. It was the double store building, on the south side of Comanche Street, that has been torn down. Thomas B. Lucas was the postmaster from 1897 to 1901. Jessie Beene (left) and Marietta Frost stand in front of the post office located in a different spot on Comanche Street. Marietta Frost was the postmistress from 1908 to 1911. Helen Frost Hawbaker gave the photograph on the left to the museum.

The first building on the right was a post office in Shabbona, Illinois. Next door was Tom Graham's Meat Market. Tom and Viola Graham bought the business in 1943 from Theodore Altringer and sold it to the DeKalb Bank in 1978. This old structure was torn down, and a new bank building took its place. It is now the home of the Shabbona-Lee-Rollo Historical Museum.

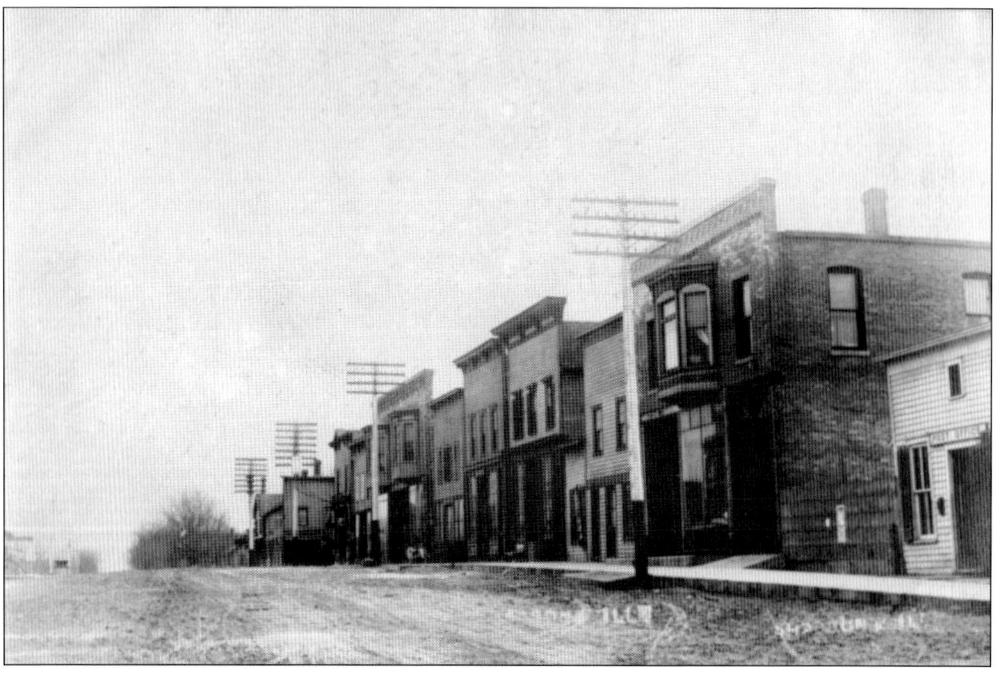

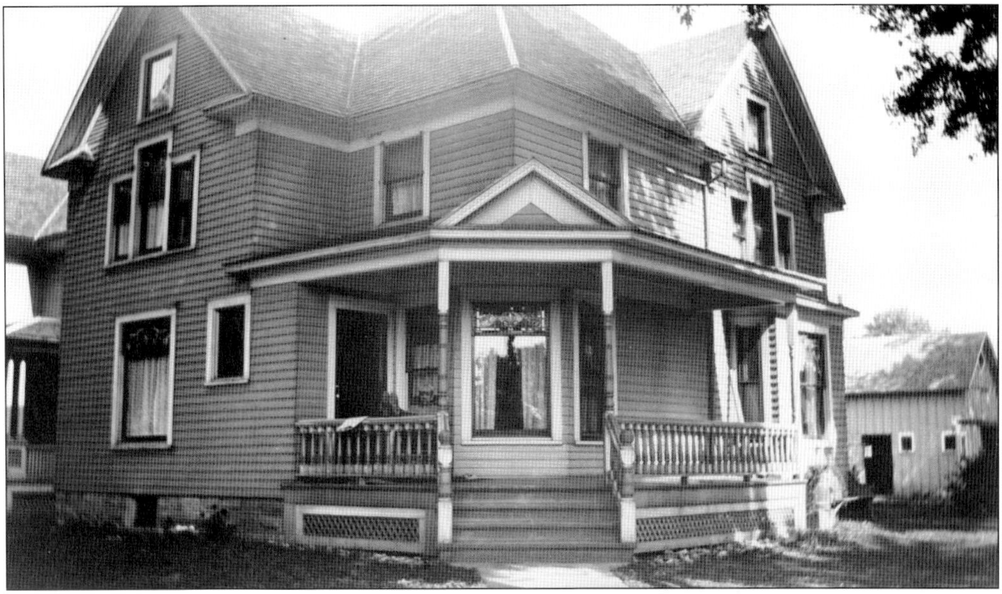

Vernon Baie was born in Clinton Township in 1908. He married Vivian Pfeiffer in 1931. They came to Shabbona in 1931 and opened a bakery. Vernon sold the business to John Schimmer in 1938 and was appointed postmaster in 1935. In 1948, he transferred to the rural route and served for many years. He and his family lived for many years in his home built in 1905 at 201 North Illini Street in Shabbona.

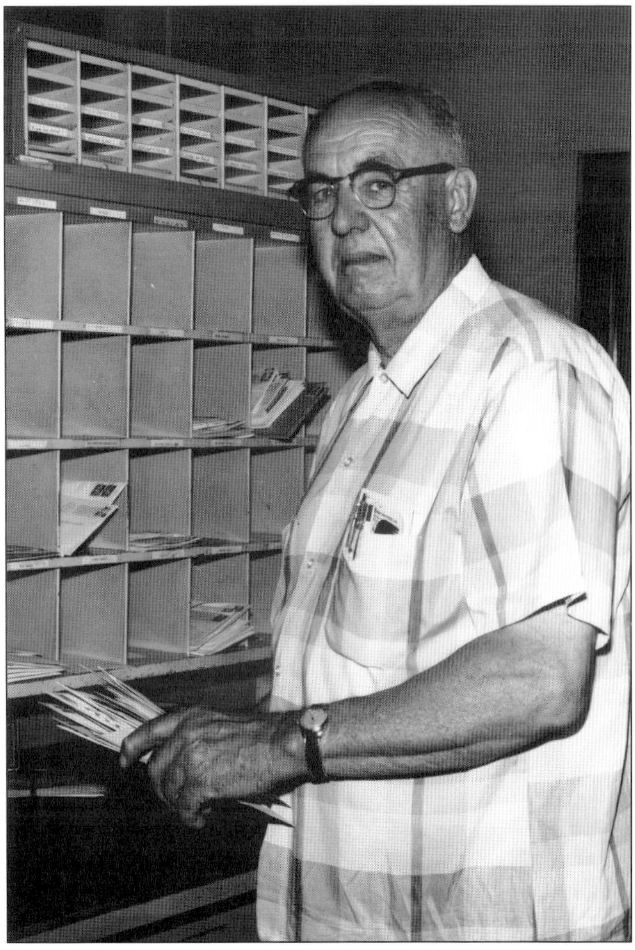

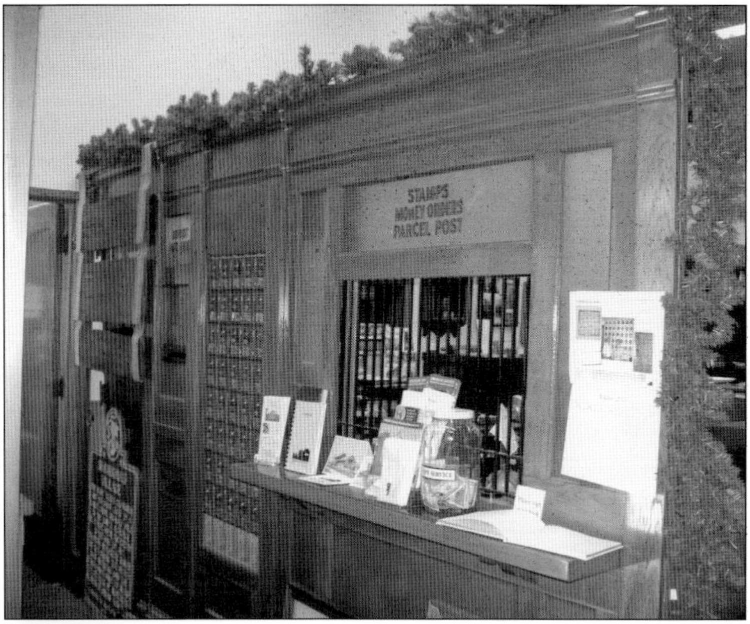

This photograph is of the window from a later post office site that is now on display in the Shabbona-Lee-Rollo Historical Museum, complete with ticket window and mailboxes. The current Shabbona Post Office, built in 1955, is located at 206 West Comanche Street, with Tim Young as the postmaster.

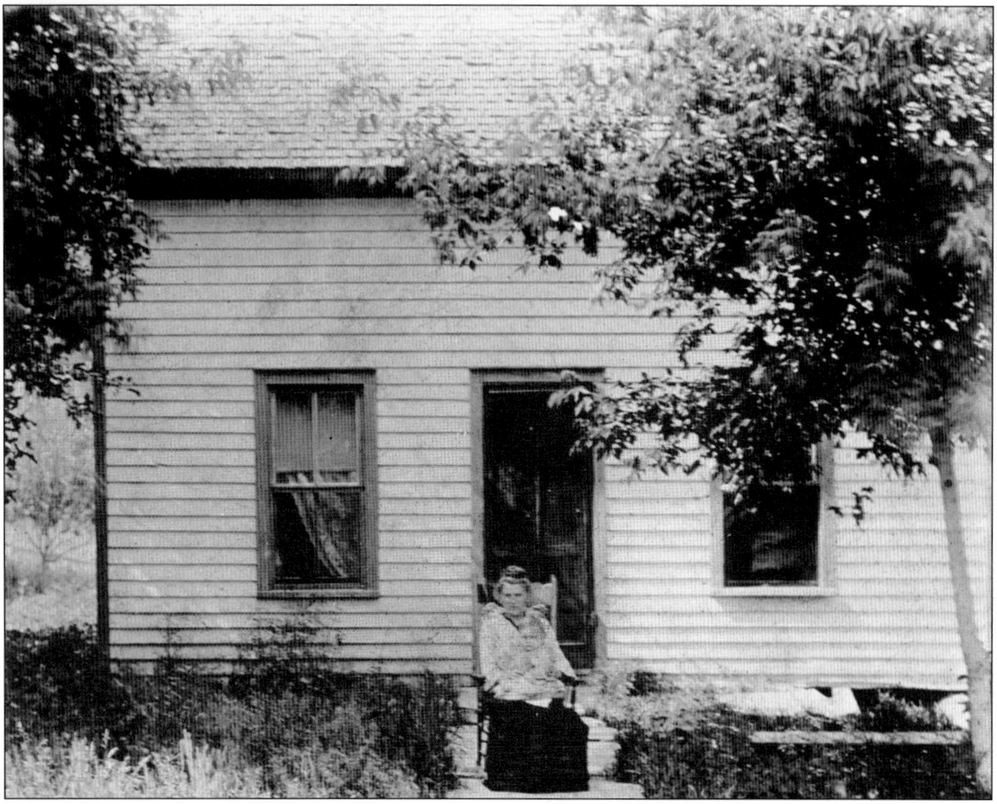

Pauline Johnson Vandre sits in front of her home on East Navaho Street in Shabbona. She was born on November 21, 1879, in Milan Township. In 1880, she moved to Shabbona and was married to Herbert Vandre on June 29, 1914. Pauline Vandre was the mother of Fritz Vandre, who was the rural mail carrier in Shabbona for over 40 years. (Courtesy of Barb Bend.)

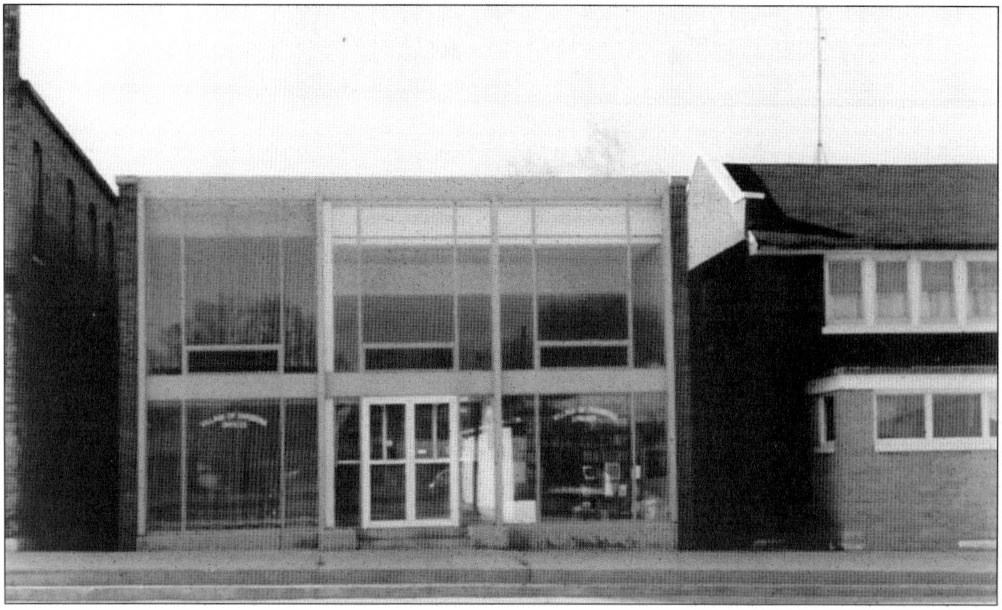

The first bank in Shabbona was owned and operated by William S. Jackson. The bank was first located in a small frame building on the main street east of the present building used by the Shabbona-Lee-Rollo Historical Museum. A bowling alley was under the old bank. A new vault was ordered for the bank but did not fit through the front door. The directors decided to erect a new building, shown above, around the vault in 1924 at 119 West Comanche Street. The business grew until it turned into a state bank in 1905, becoming the Farmers and Traders State Bank. A new bank, pictured below, was built in 1974 at 203 West Comanche Street in Shabbona, Illinois. In 2009, a branch bank was opened at the corner of Routes 23 and 30, east of Waterman, Illinois.

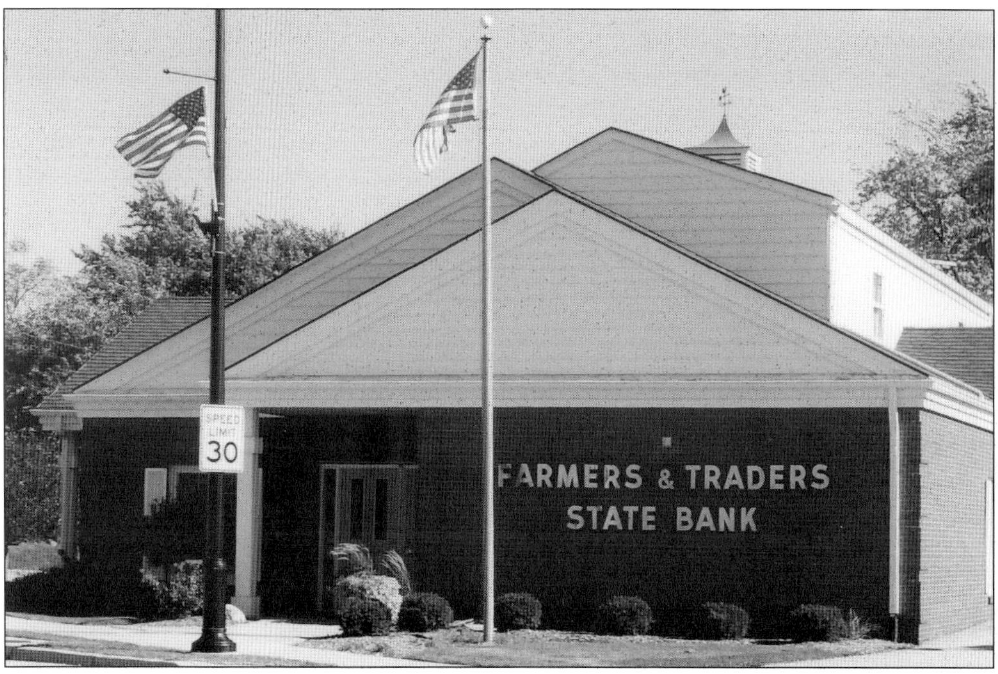

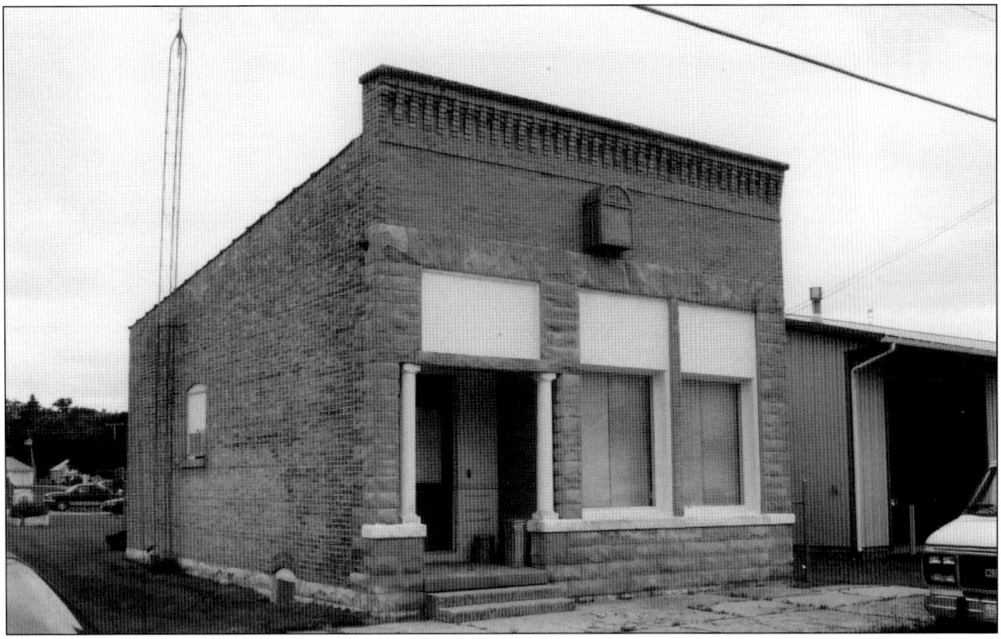

The Lee Bank Building was well equipped with modern fixtures, including safety vaults and one of the best fireproof and burglarproof safes of its kind. The bank was closed by moratorium in 1931. Martin Botten purchased the building. Later, it was sold to Mary Kennedy, and the post office was moved into it. Mary, granddaughter of one of the founders of Lee, was postmistress and retired after serving the community for 40 years. The structure is now an apartment building. The current Lee Post Office, seen below, is located in a new building on the south side of town.

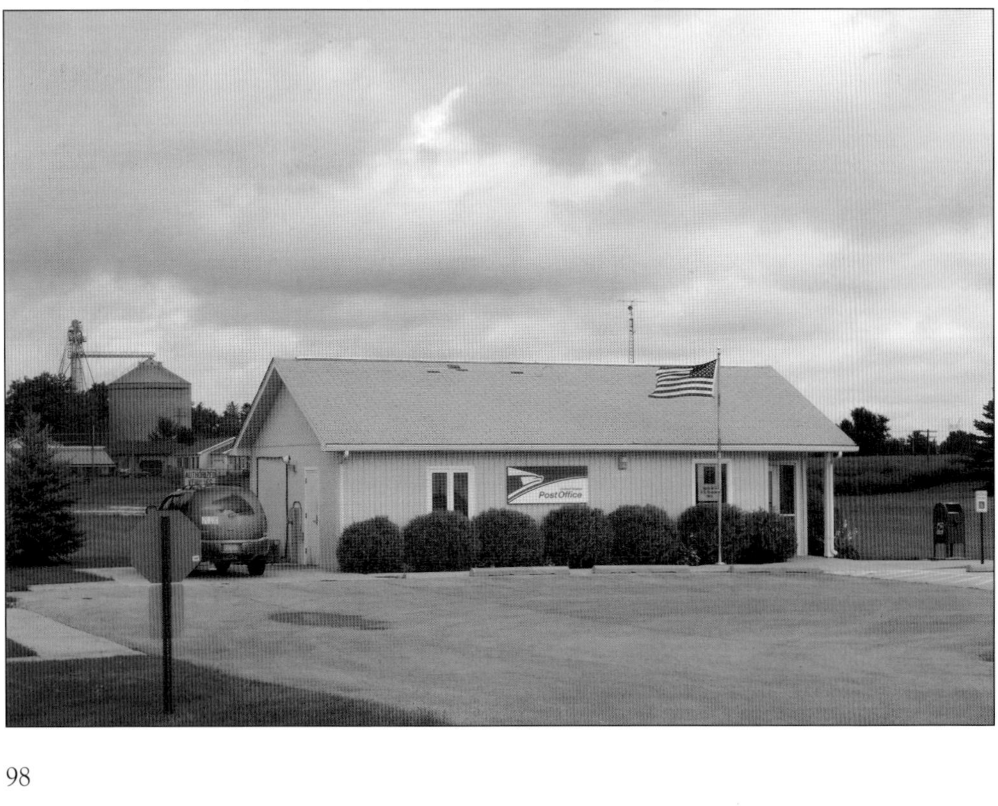

Eight

Farming across the Area

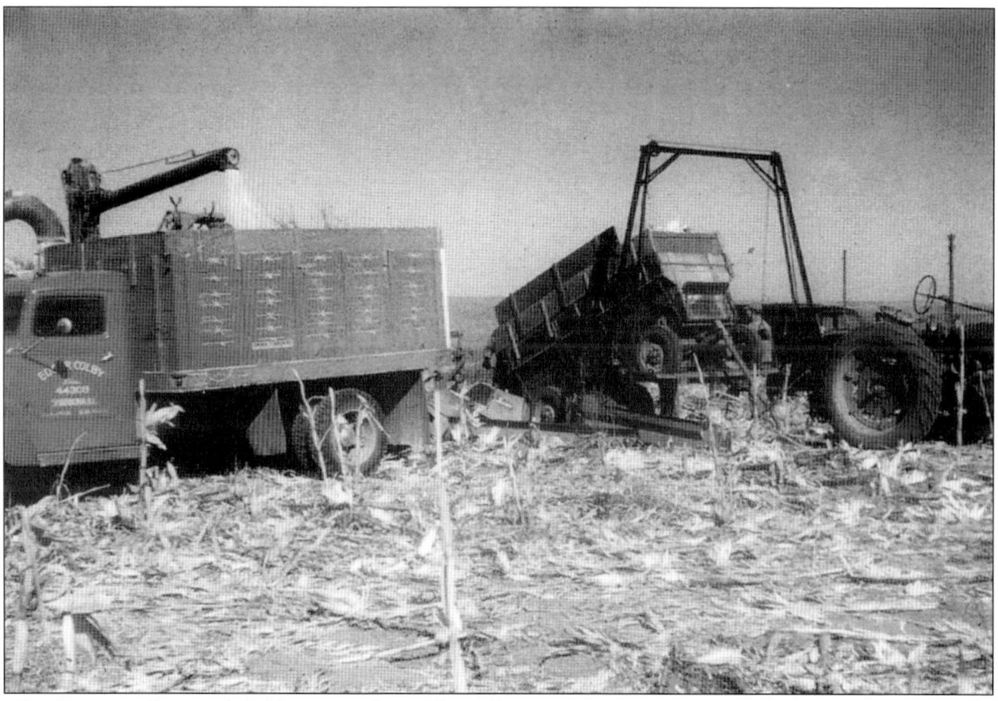

The farmers who settled Illinois in the early 19th century were mostly of English, Irish, Scottish, and German descent. The completion of the Erie Canal in 1825, the passage of the Homestead Act in 1862, and the invention of barbed wire in 1874 by Joseph Glidden all contributed to the disappearance of the open frontier. Edgar Colby of Shabbona had a corn-shelling business for many years in the area.

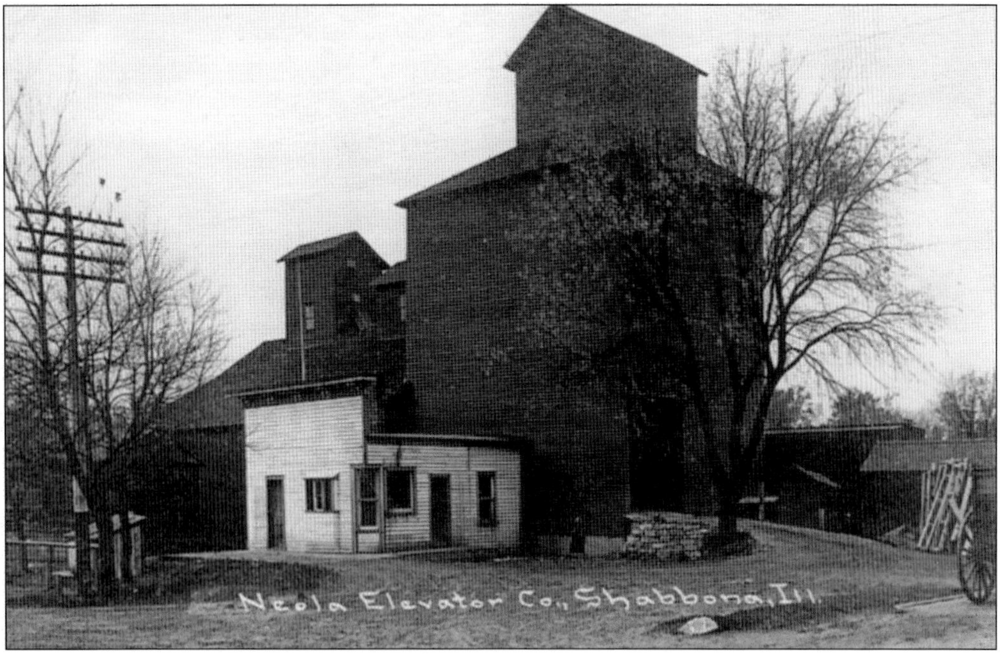

In the spring of 1872, Smith and Pollack operated an elevator in Shabbona. They elevated grain by means of a blind horse that walked in a circle to push the machinery. Griffin and Todd had a hardware, coal, and lumber shed on the west side. In the early 1900s, A. Larson, an agent, ran the Neola Elevator. In 1932, it was sold to a Mr. Herrmann of Scarboro.

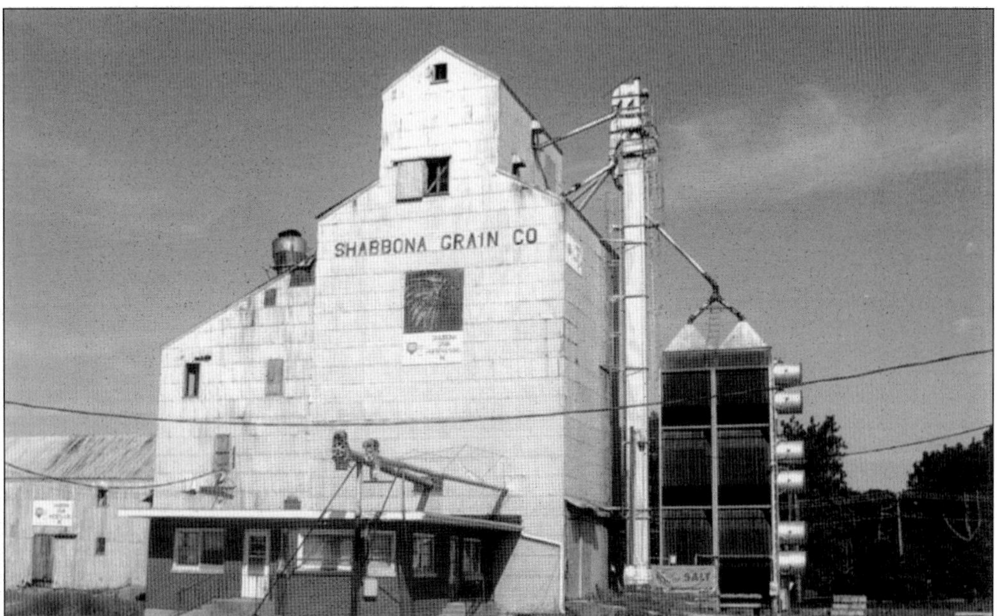

Samuel Bouslough was one of the first to establish an elevator with grain to be loaded and unloaded by hand. The elevator was eventually sold to Alexander Lumber Company in 1925, and L.S. Thorsen was made manager. After 20 years, Thorsen retired, and Harlan Douglas managed the business. Hueber Grain owned the elevator when it closed in 2009.

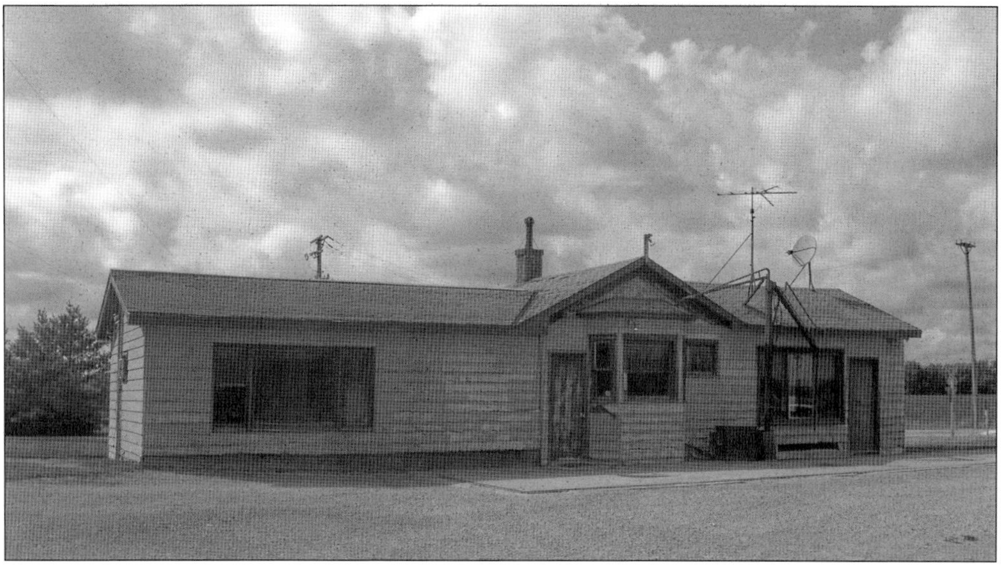

Strong and Strong built a new elevator in Rollo in 1923 after the previous elevator burned. The new building was 24 by 36 feet in size, over 70 feet high, and built of flat timbers spiked together. The building was equipped with the latest automatic dumps and elevating machinery and was up to date in every way. The office building (above) remains. The old buildings have been torn down, and Mullins Grain now owns the business. Mullins Grain Company, pictured below, is a private company that was established in 1969, with Robert Mullins as owner. The company has three sites in Lee, Rollo, and Shabbona.

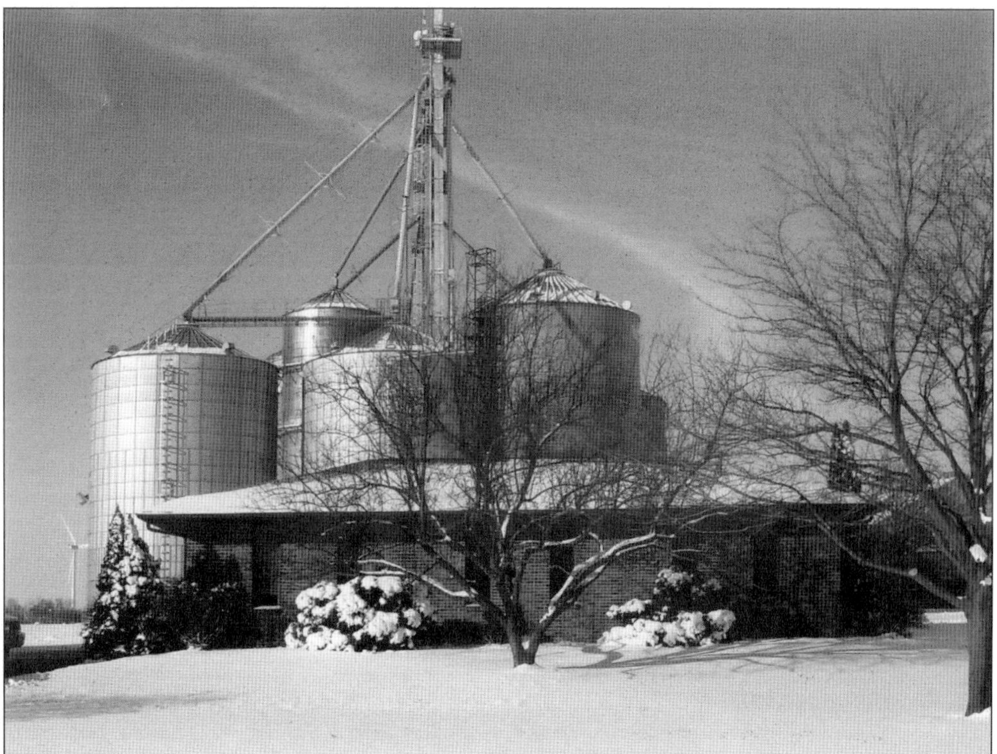

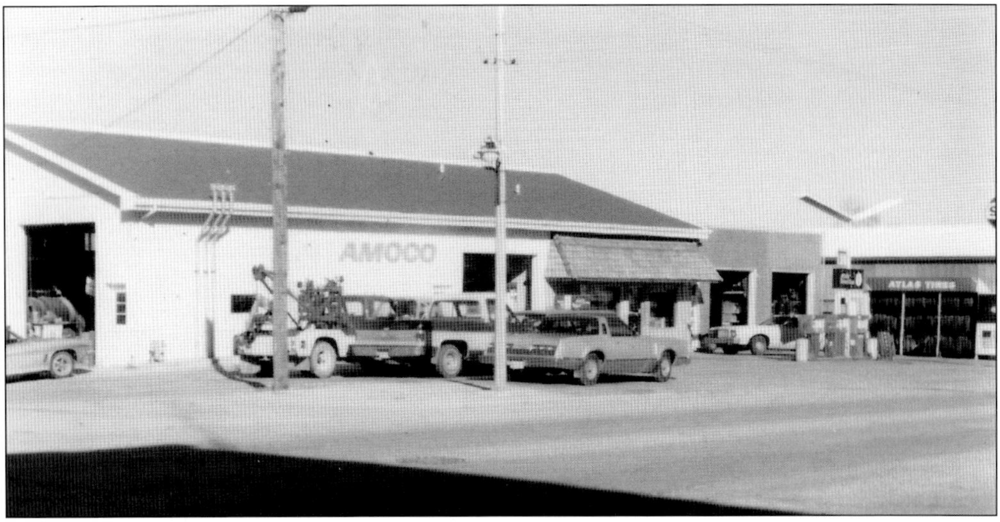

After 81 years, the old Swanson Building on the northeast corner of Illini and Comanche Streets in Shabbona came down in 1971 to make way for Falk's new Standard Oil Building. It was best known as the general merchandise store operated by the late Albert Larson. The picture shows Falk's Standard Oil as it looked around 1971. In 2009, the building burned and is now an empty lot.

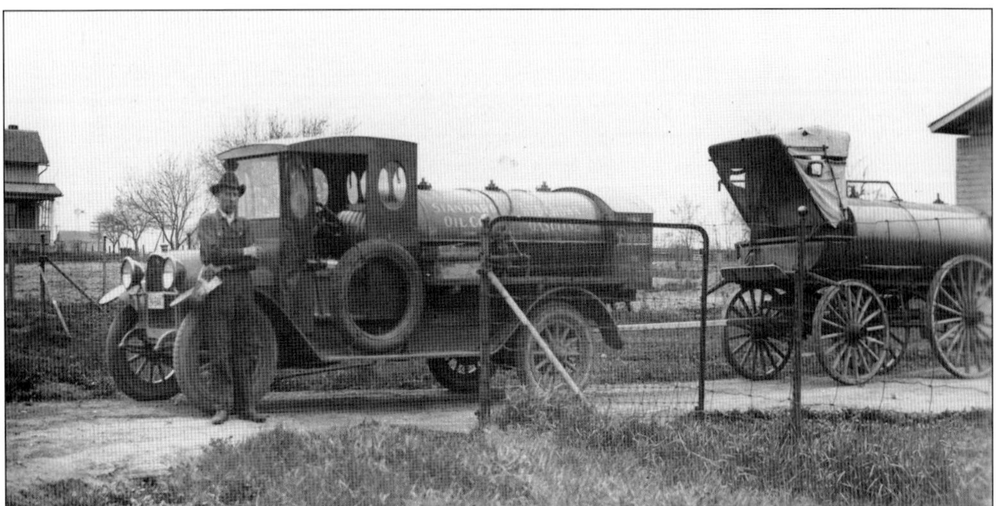

LeRoy Todd got his start with the Standard Oil business through his father, William, in 1933. In 1951, he took over the business because of his father's ill health and served the community until 1970. He then sold to Jerry Leifheit. Red Crown, one of the popular types of gasoline in the 1920s, was often delivered in this Standard Oil truck, a forerunner of a present-day tank truck.

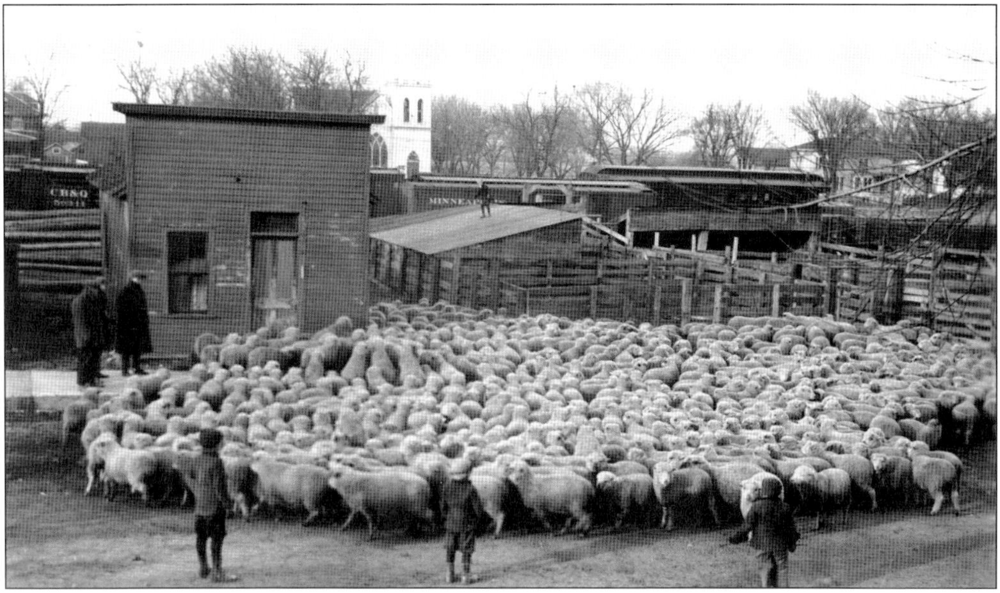

The farmers brought their sheep into Shabbona for sale. Getting food animals to market required herds to be driven to railheads, where they were loaded into stock cars and transported eastward to regional processing centers. The stockyards were located on the north side of the main street in Shabbona. In this image donated by Fritz Vandre, note the children surrounding the sheep that belonged to C.P. O'Kane. The United Church of Christ is in the background.

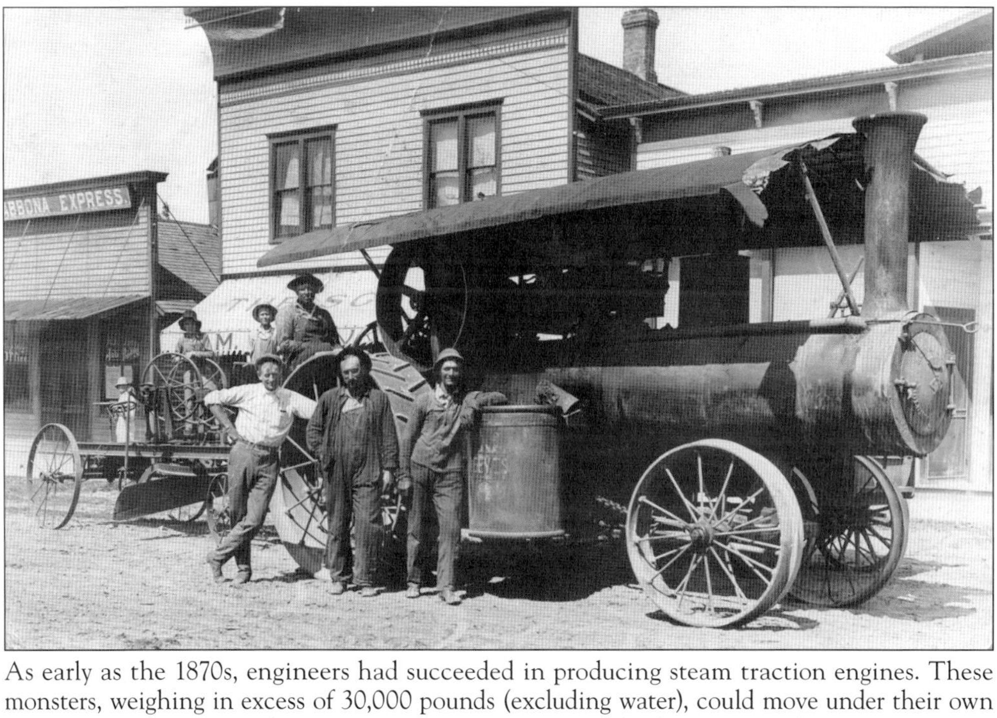

As early as the 1870s, engineers had succeeded in producing steam traction engines. These monsters, weighing in excess of 30,000 pounds (excluding water), could move under their own power and had impressive horsepower capacity. Unfortunately, their size, mechanical complexity, and the constant danger of explosion made these traction engines unusable for most farms in North America. The Shrader brothers owned this 1913 machine.

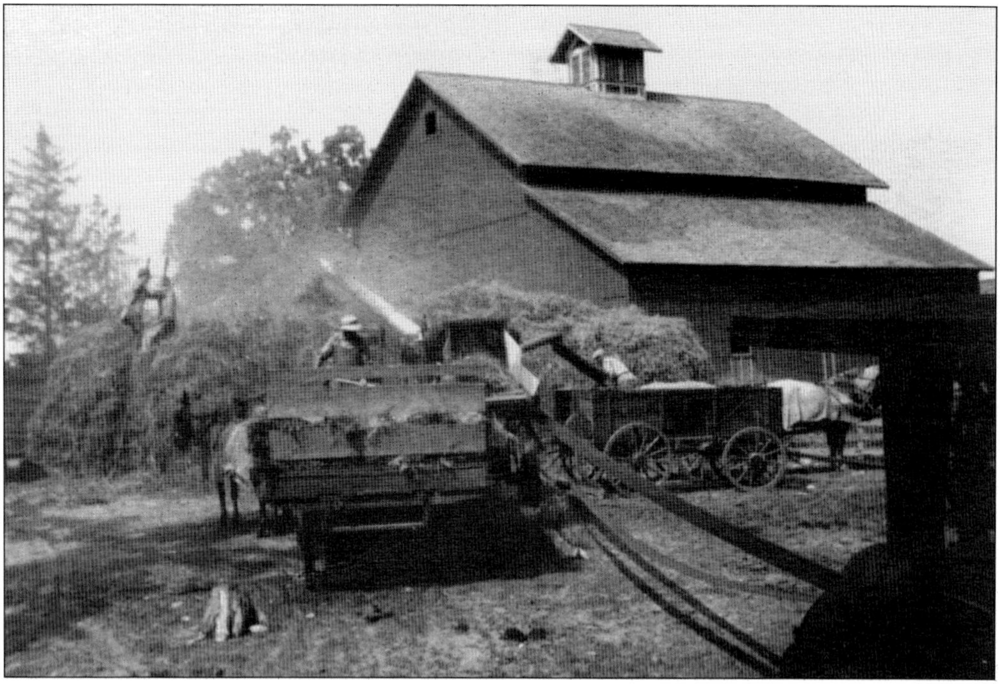

In the 1800s, the Stevens family was threshing oats. The oats, bound into bundles and stacked, were brought to the grain separator powered by a horse-drawn portable steam engine. The straw was then blown into the loft of a barn for later use. The horses ate the oats, and straw separated from the oats was then used as bedding for farm animals.

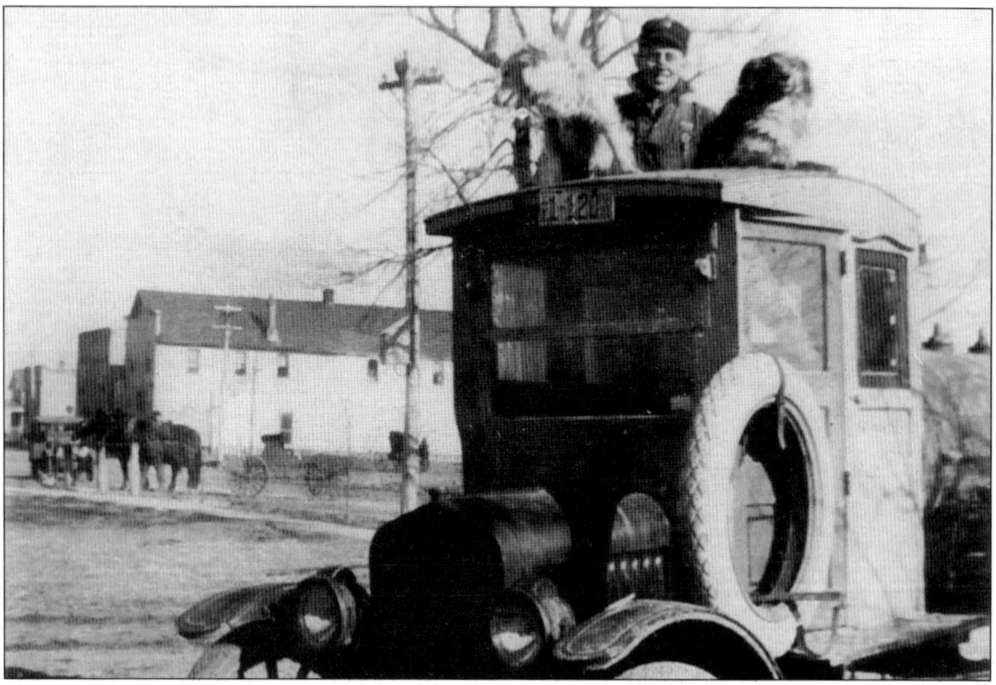

Ellis "Punch" Rees stands between Shep and friend atop his gasoline delivery truck south of Phil Schoenholz's Grocery and Dry Goods Store in Scarboro, Illinois. (Courtesy of Shirley Rees.)

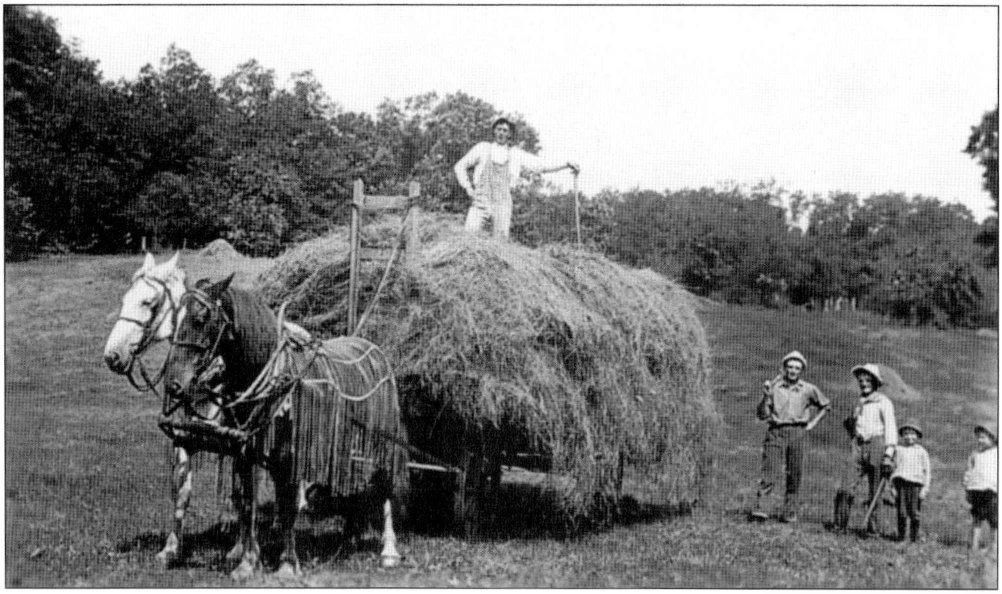

The farmer on the pile of hay is Ray Stevens. The photograph of the Stevens farm, donated by Rita Stevens Blanchard, was located at the southeast corner of Preserve and Shabbona Roads. It is now part of the Shabbona Lake State Park.

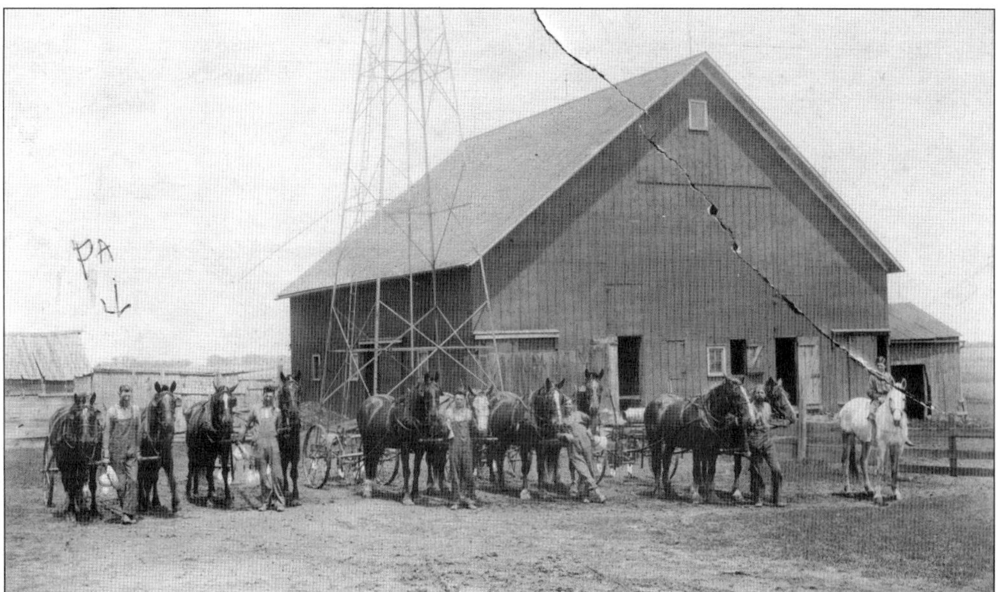

The Duffy farm is located east of Shabbona on Duffy Road. In this photograph donated by Helen Modeen, each man in the family—from left to right, James (father), Martin, Laurence, Sylvester, and Henry—had a team of horses. Agnes Duffy is shown on a white horse. The Buchholz family now owns the farm.

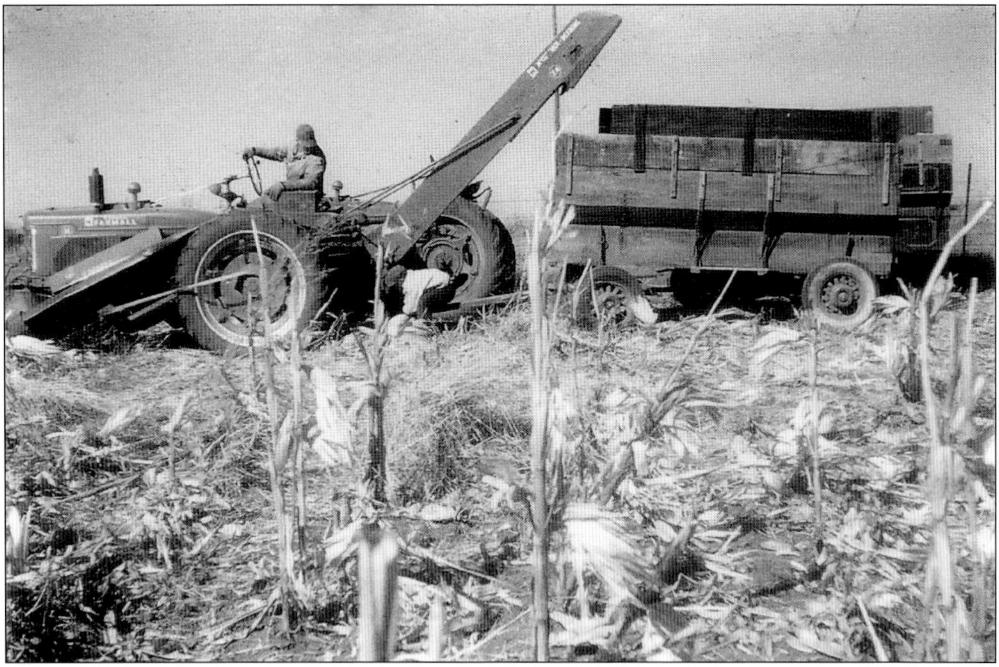

The Lyle Jacobson farm was located on Route 30 southwest of Lee. This image shows Emory Thompson picking corn in the late 1940s or early 1950s. He is using an International Harvester Farmall tractor with a two-row, mounted corn picker. Notice that the open cab is not covered from the elements and that the small child is helping to hook up the wagon.

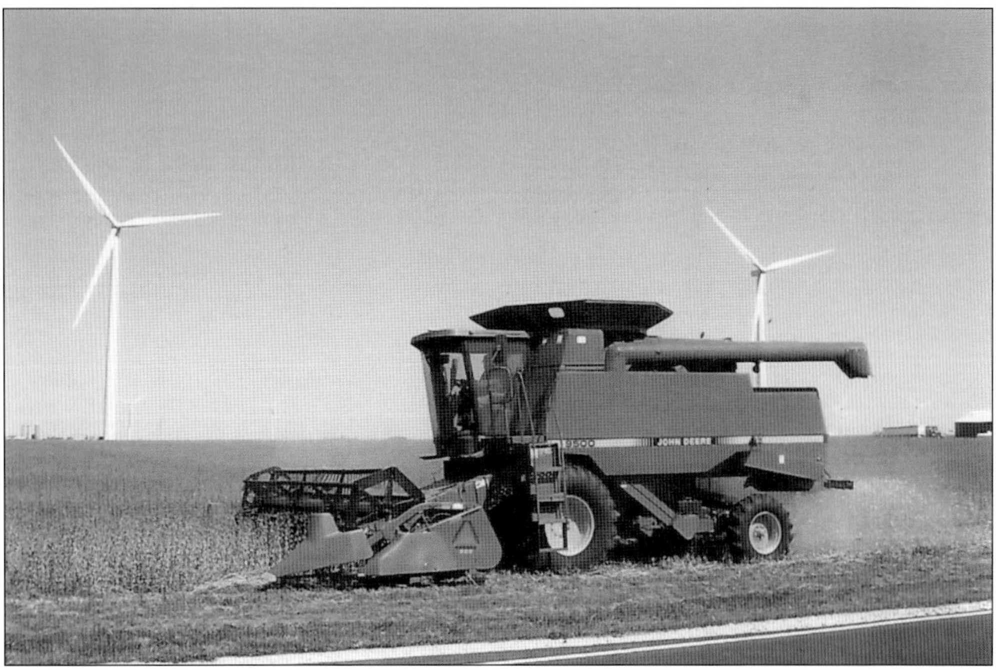

An example of present-day farming in 2010 is Joe Barshinger on Lee Road in Lee, Illinois. He is combining beans with a John Deere 9500 tractor and combine. Barshinger is secure inside his cab. Notice the windmills in the background.

Nine

Interesting and Fun Sites

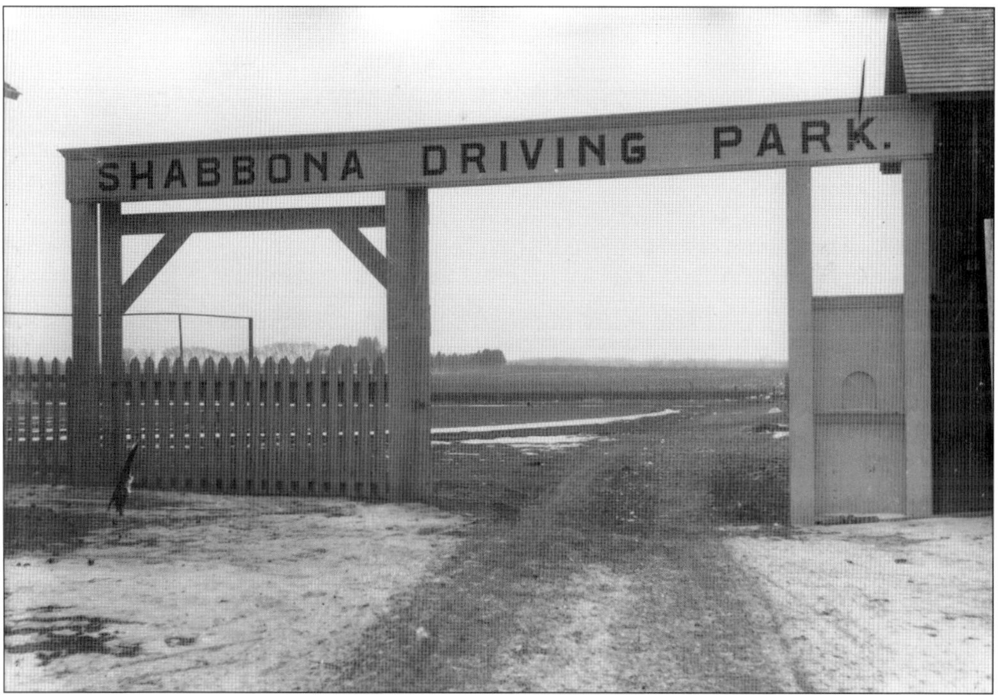

The area covered by the Shabbona-Lee-Rollo Historical Museum is large, including parts of DeKalb, Lee, and LaSalle Counties. One does not have to look far to find curiosities, landmarks, and mysteries from the past. For entertainment, there was a racetrack on the acres owned by the Shraders, where local fairs and events were celebrated. The site is now Apache Street in Shabbona.

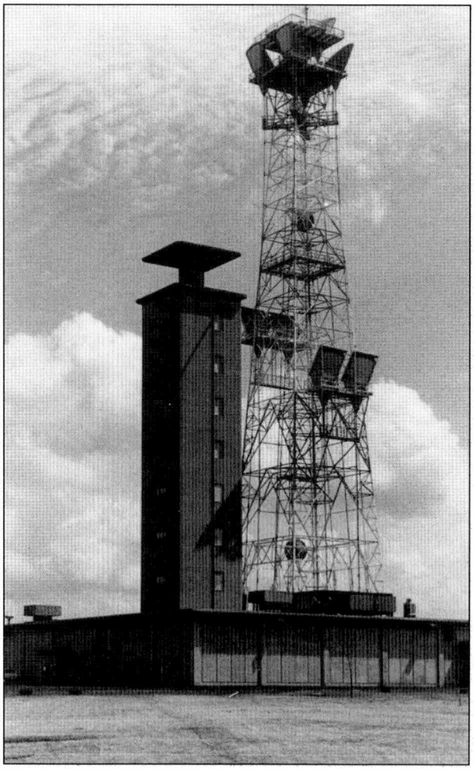

American Telephone and Telegraph, Inc., built a structure to house modern power equipment, solid-state switches, and circuitry at its microwave tower near Lee in 1950. Born in 1888 in Lee, Illinois, of Norwegian emigrants, Randolph Eide worked in his father Tollef Torris Eide's general store. He enrolled at the University of Illinois and earned a BA in 1910. In 1930, he became president of Ohio Bell Telephone. The American Telephone and Telegraph Tower outside of Lee, Illinois, was dedicated to him in May 1949. By 1993, all service had been turned off at the Lee site, and Western Electric crews removed all equipment inside the main building. The site was purchased in 2002, and since then, the owner has been performing historically accurate restoration of the concrete tower.

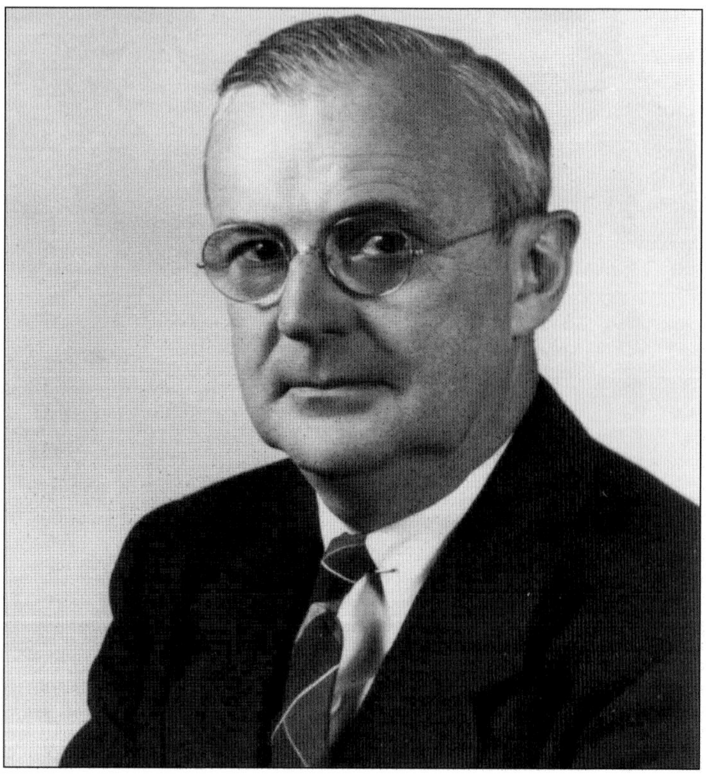

Ross Grove Cemetery is located half a mile north of Chicago Road and was deeded in June 1845. It is a wooded area on the west side of the road, south of Big Indian Creek. The first church in Paw Paw Township was built at Ross Grove by the United Presbyterians in 1861. The building was moved to the Shepardson farm, now owned by Gordon Johnson, and used as a barn until it was torn down. This cemetery serves as a final resting place for an unknown number of residents. Of the remaining headstones in the bramble-covered woods, members of the Harper and Marshall families, buried prior to 1893, were found by members of the Shabbona-Lee-Rollo Historical Museum. Shown below is a group firing a cannon in Shabbona Village Park in 1897. Notice the United Church of Christ in the background.

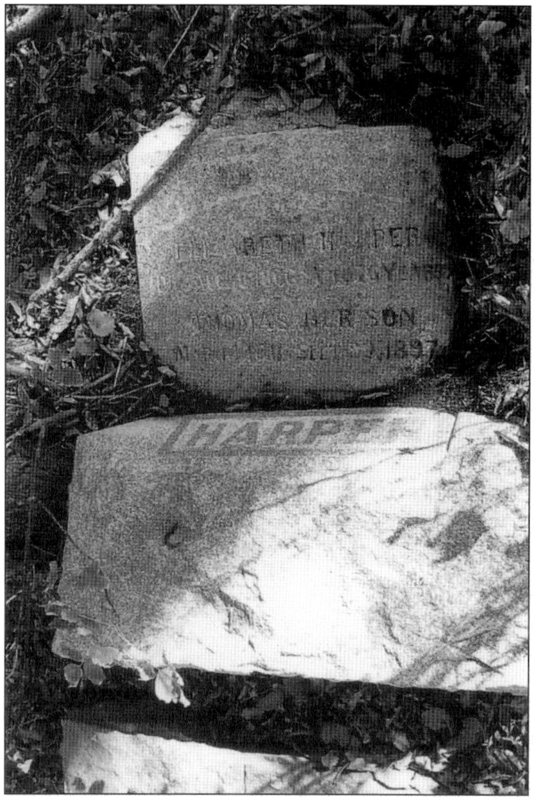

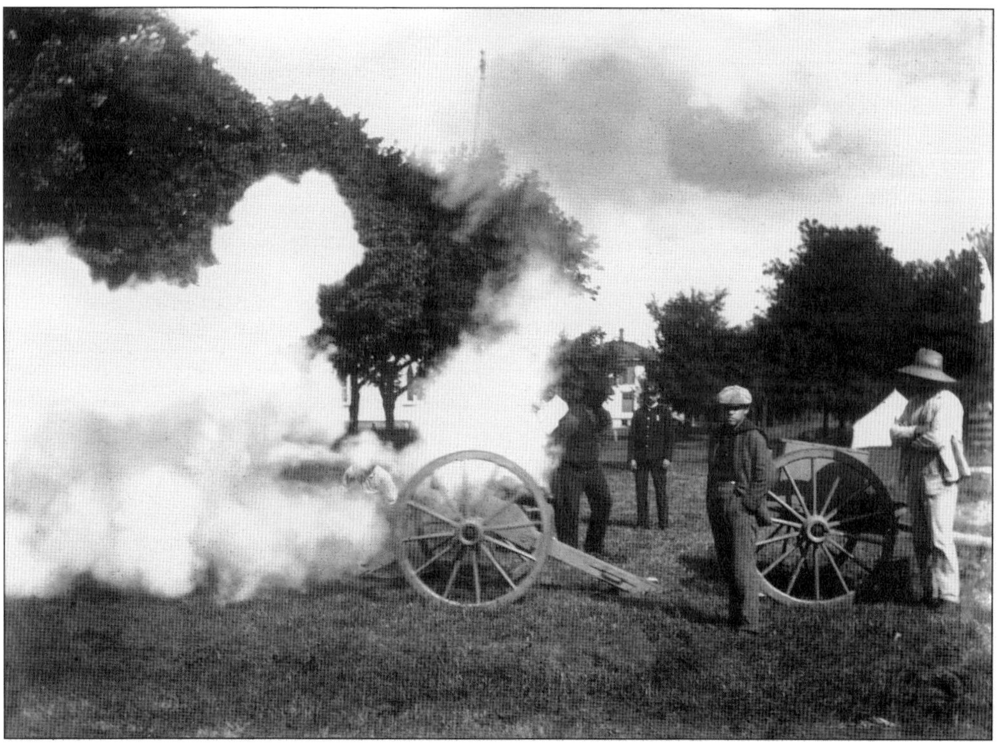

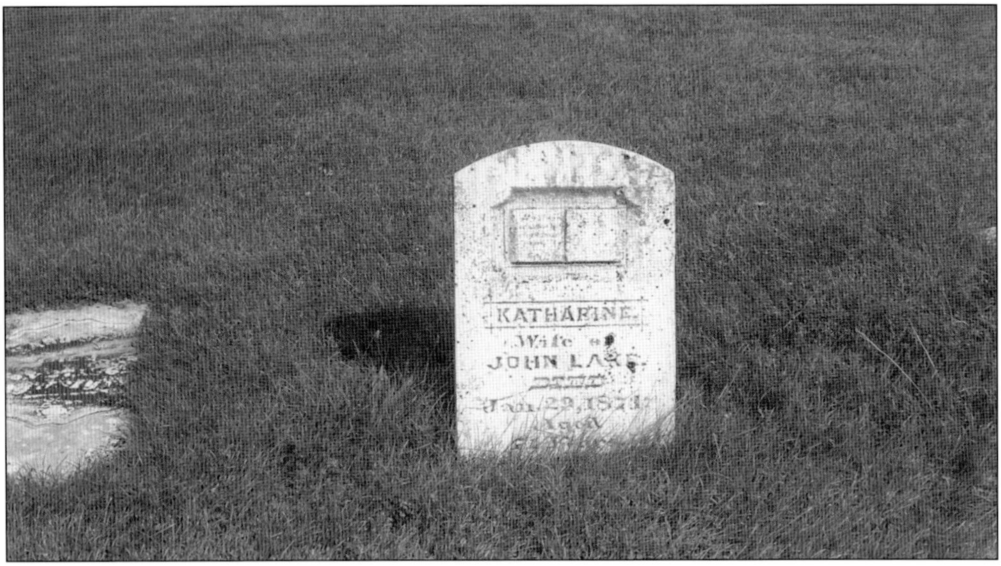

"He should have tested a daisy first-before he ever tried to squire me. Katrina Loves Me; Katrina Loves Me Not-NOT!!!" wrote Washington Irving in the "Legend of Sleepy Hollow." Katrina's character was based upon Katharine Van Tassel Lake. Irving was a family friend and found her family interesting subjects for his tale of Ichabod, Brom Bones, and the Headless Horseman. Katharine was born in Albany County, New York, on January 1, 1802. She married John Lake in New York and moved with her family, which included eight children, to Hancock County, Illinois, where her father had bought a farm and had moved there some years earlier. Her son Almon Lake bought land in Paw Paw Township, and that is where she died on January 29, 1874. She was laid to rest in Stevens Cemetery, located southeast of Rollo, Illinois.

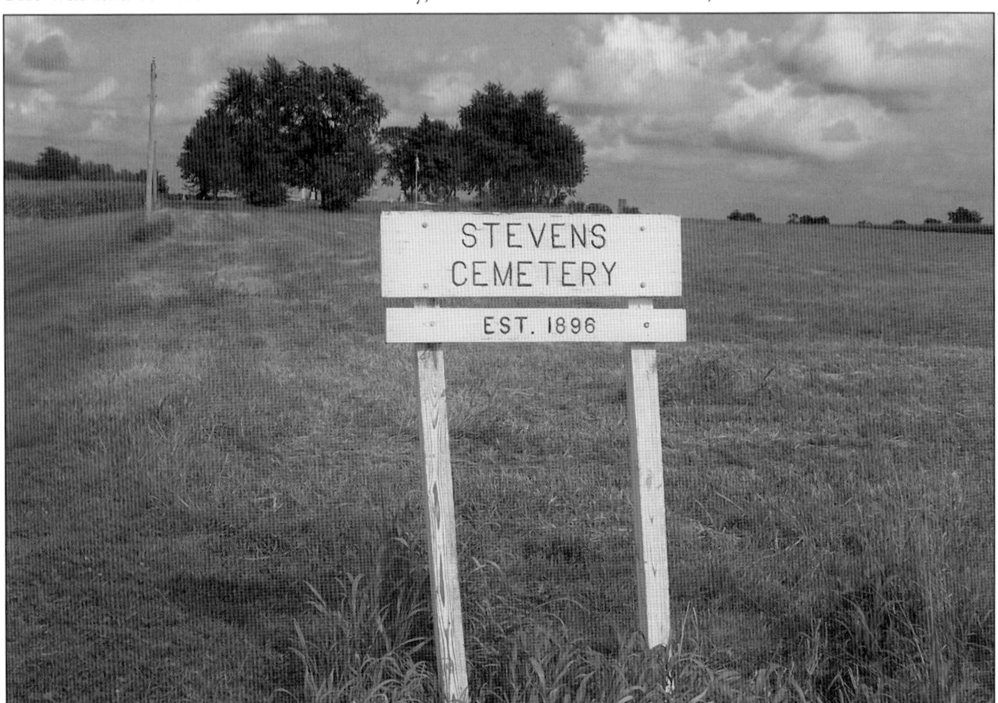

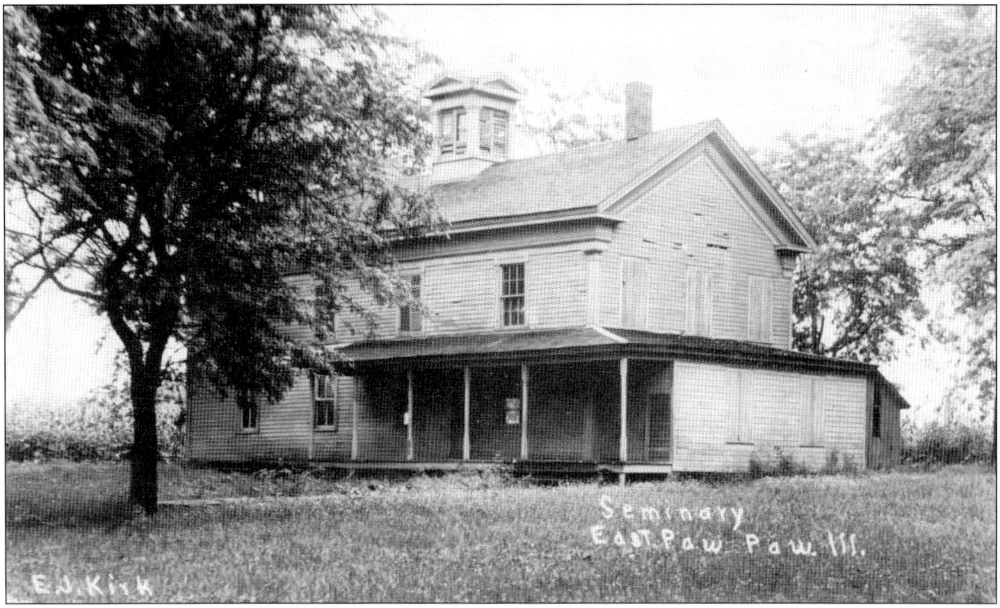

A school of local prominence was the East Paw Paw Teacher's Institute and Classical Seminary at the corner of Chicago and County Line Roads. It was an educational forerunner for children of pioneer settlers. The school was plagued by fires, being taken down in October 1913. One of the best-known graduates was Clinton Rosette, former editor of the *DeKalb Chronicle*, who was born and raised on a Paw Paw Township farm in 1850. The only tangible asset of the seminary is the old school bell, which hangs in the steeple of the Rollo Congregational Church. The inscription says, "Teachers Institute and Classical Seminary, East Paw Paw, Illinois, 1882, *Lumen Petimus*, Light We Seek." Pictured are students of the Rollo High School class of 1930, which included a member of the Rosette family.

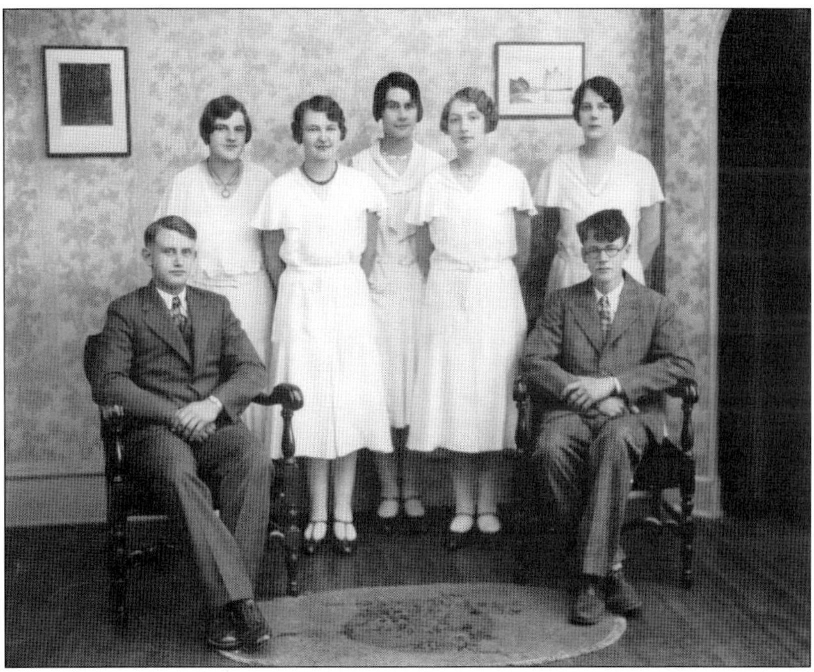

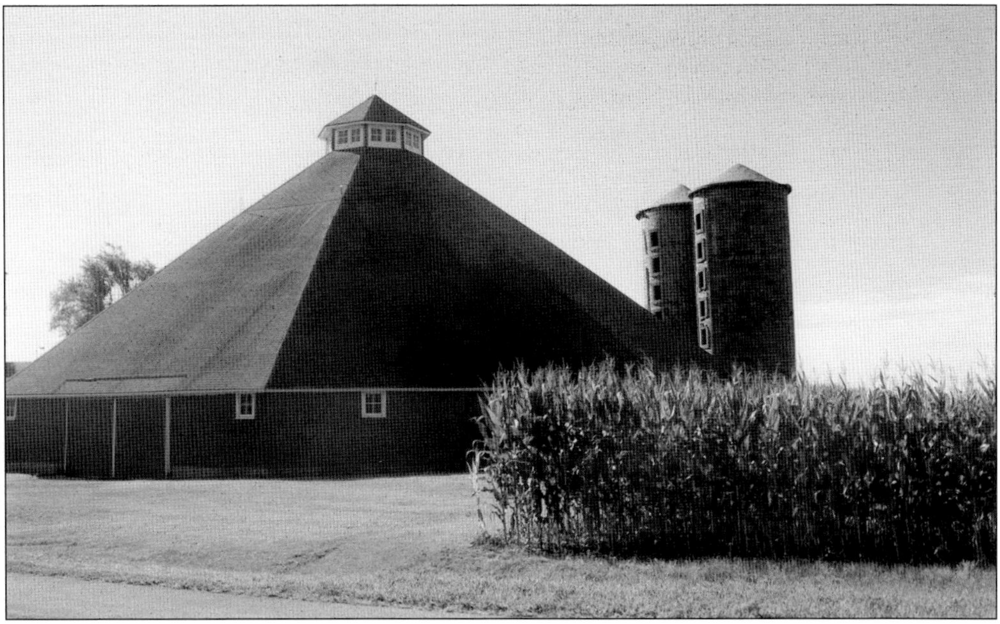

William Swanson assisted the Svihus brothers in building this red, octagon-shaped barn in 1908 on South Paw Paw Road for William M. Goble. The three-story, eight-sided barn found west of Rollo cost $3,714.70 to build. It is made of 3-by-12-inch boards 34 feet long and 18-by-22-inch boards, which were 30 to 35 feet long. The story has been told that 10 teams of horses and their wagons could stand in a circle on the inside of the outside wall. Pulleys were used to move the machinery up to the other floors. The unique barn is still standing. Shown below is an early corncrib located on a farm west of Shabbona. Farmers in the 1930s used crib structures that had open ventilation for air-drying crops of ear corn.

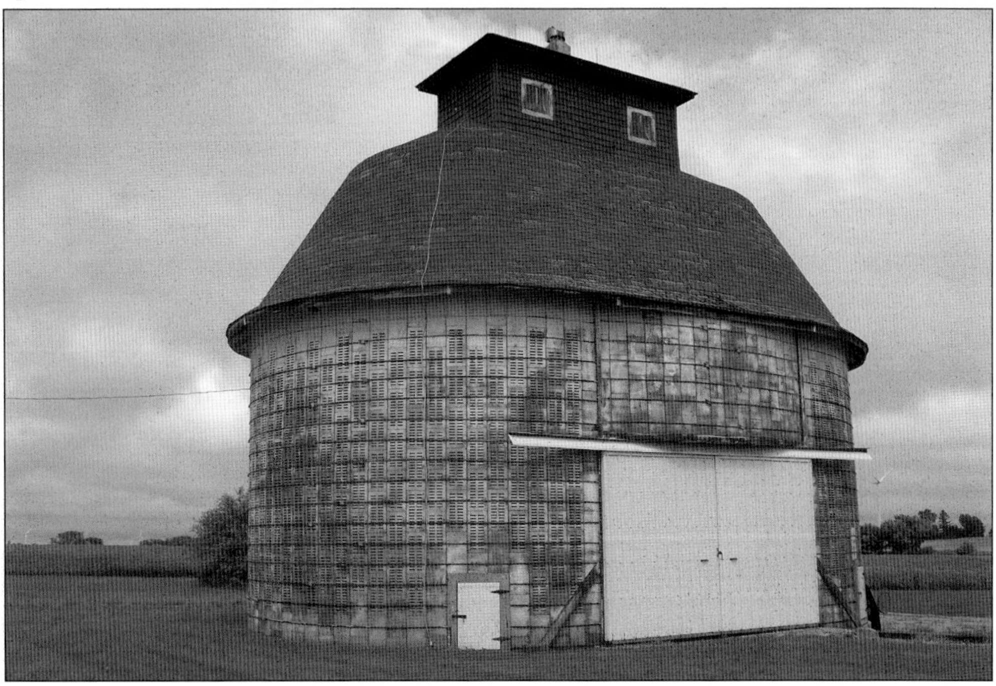

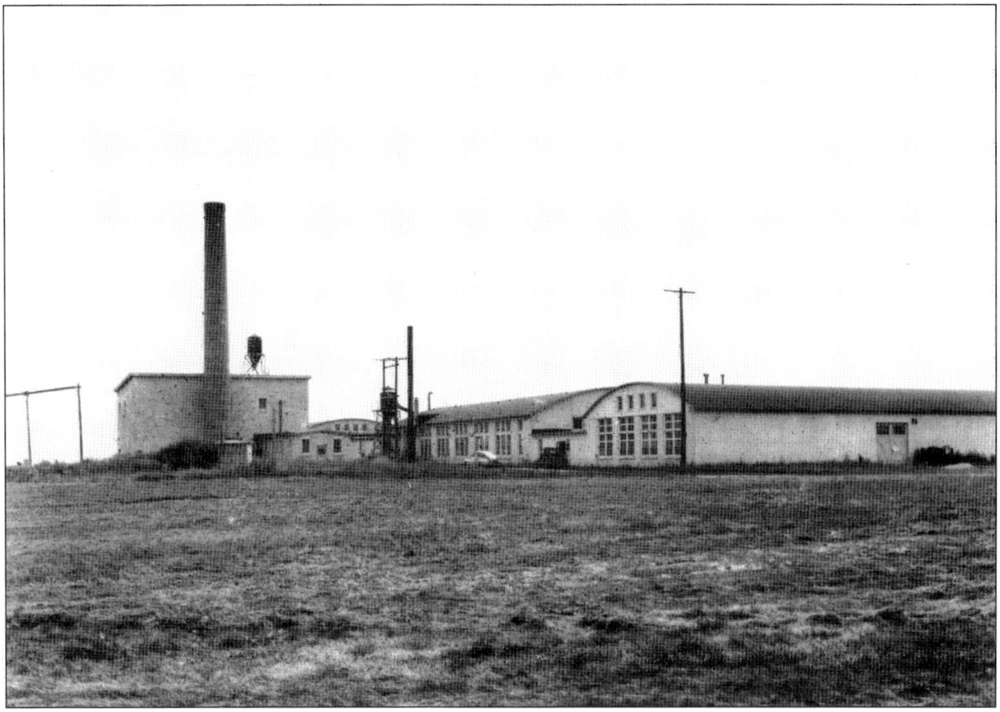

During World War II, the government built a plant for the production of cordage used for all types of war needs. The growing of hemp would only be a temporary proposition. In February 1943, it was decided to build a plant near Shabbona for $350,000 with an annual payroll of $200,000. Hemp was grown, and the plant was built in 120 days, with E.E. Houghtby as plant manager. The factory employed as many as 100 people. There were six buildings that housed displaced Japanese Americans who lived on-site for about two and a half years. German prisoners from Camp Grant were also brought to Shabbona to work in the fields. Work at the mill ended in 1946. A public auction of war surplus properties was held on the grounds. The picture at right is of Benny Prestegaard, the hemp plant foreman.

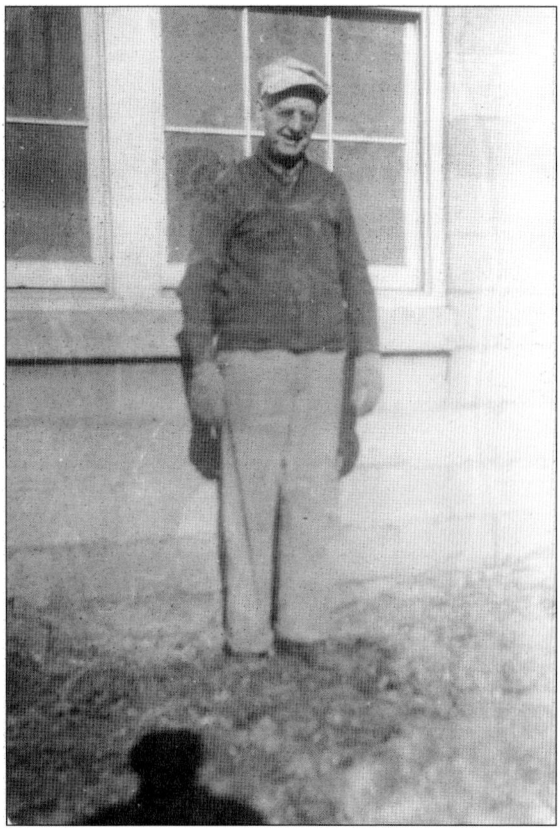

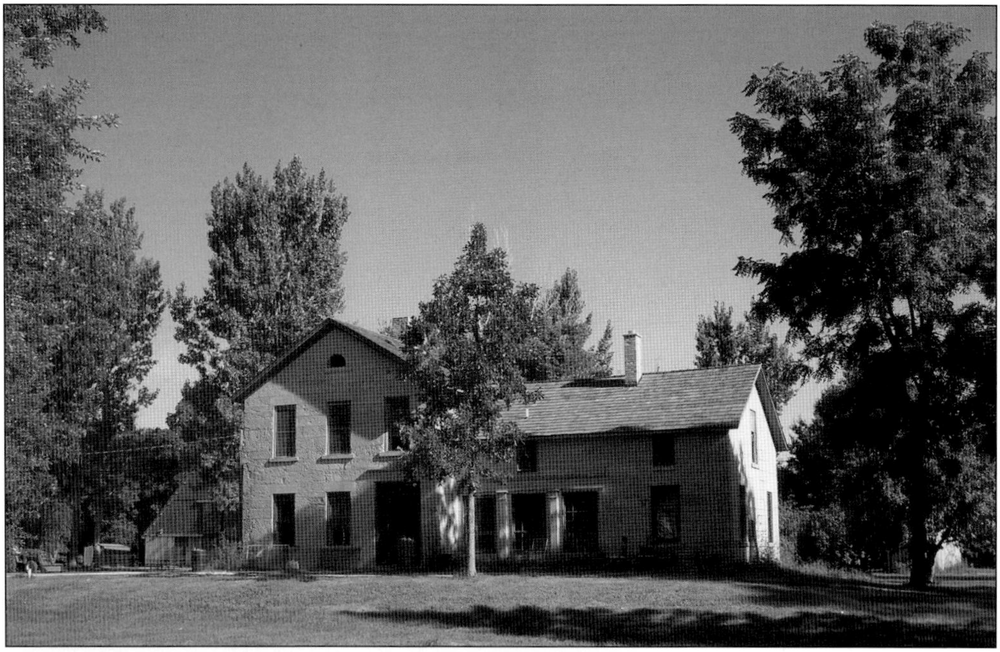

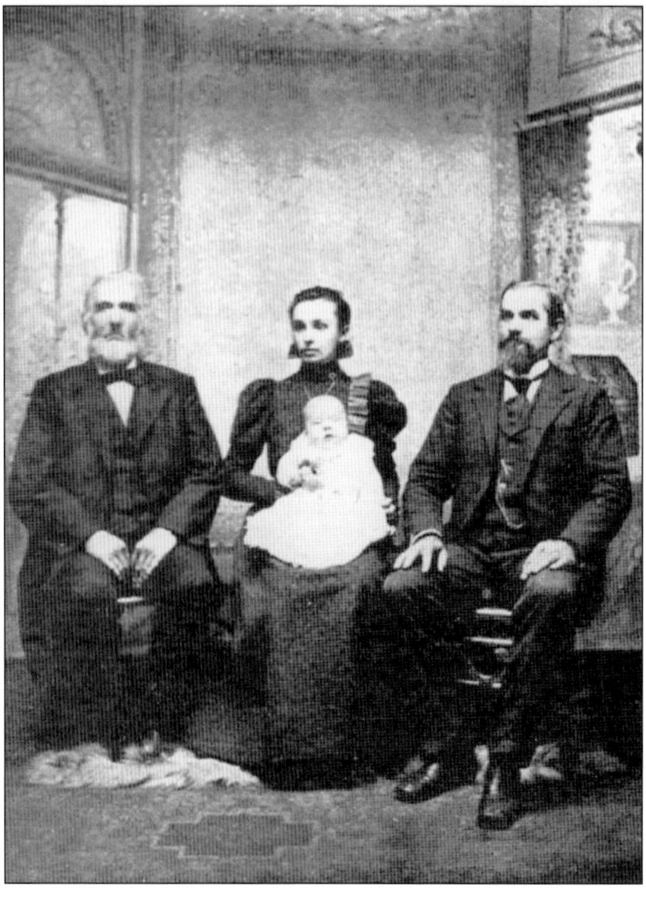

Matthew Nesbit, having only known stone houses in his native Scotland, began construction of the stone house east of Rollo in 1848, bringing the limestone blocks by oxcart from the Ottawa quarry. It took 15 years to complete the house with help from his brother Peter. The house is of Greek Revival style with approximately 2,500 square feet of living space. On the lintel is carved "M.N. 1863," the initials of the builder and the date of completion. Greg Larson of Champaign, Illinois, now owns it, and reenactments are held on the grounds. Four generations, from left to right, William, Nellie Vail (née Weddell) Bullis holding Mefflin, and C.V. Nesbit, are shown in the picture. The Weddell families were early pioneers of Paw Paw Township. Charles Weddell was married to Mary Nesbit, daughter of Matthew Nesbit.

According to the *Paw Paw Times* on June 12, 1930, there is great excitement among the farmers north of town up into Willow Creek Township relative to the possibilities of oil wells on their farms. Certain scientists have told the people that there are large deposits of oil under their farms. The oil derrick on the Eraas farm was 96 feet high, and the drill was powered with a 120-horsepower gasoline engine. The company got into legal problems and was shut down. Eventually, a well was drilled on Tower Road, which produced gas only. Pictured below are the Flying Farmers, an organization of families with a mutual interest in promotion and use of airplanes in agriculture. The group organized after World War II. Lloyd Forrer is in the center of the back row, and Lloyd Buland is second from left in the first row. (Right, courtesy of Rita Stevens.)

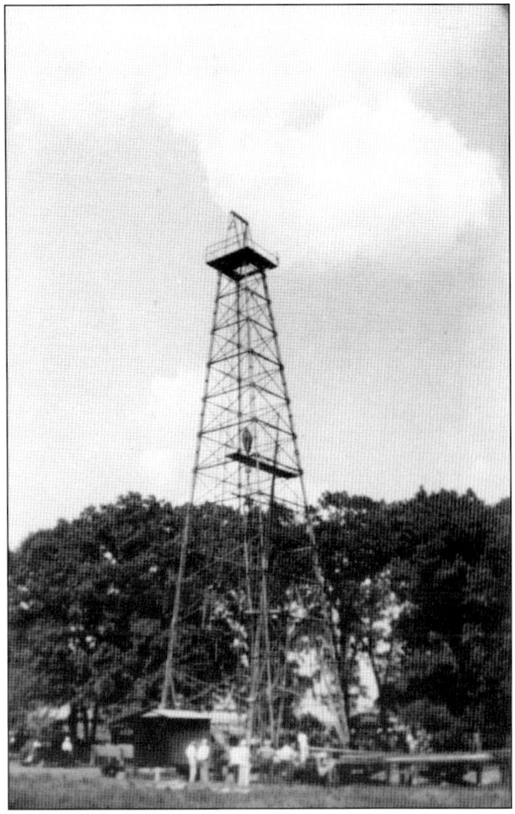

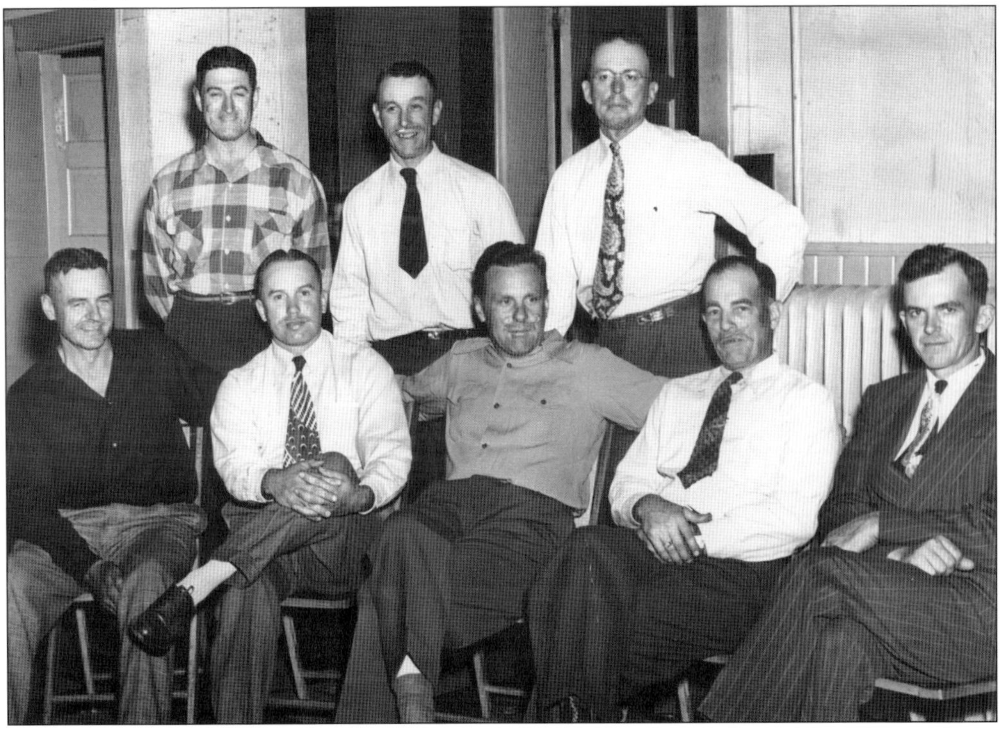

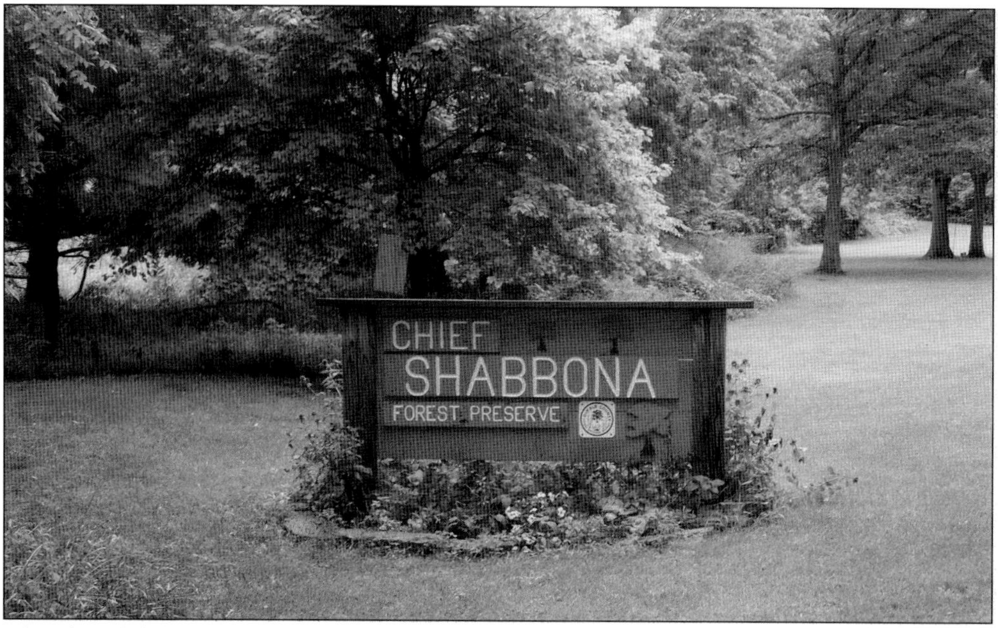

The idea of dedicating a suitable piece of his old reservation in Shabbona Township to Chief Shabbona for park purposes became a reality in 1941. The DeKalb County Forest Preserve Board bought a total of 88 acres, including 57 acres of the William Husk estate and 15 acres of the W.J. Norton estate southeast of Shabbona. This site is where Chief Shabbona lived and three of his children were born.

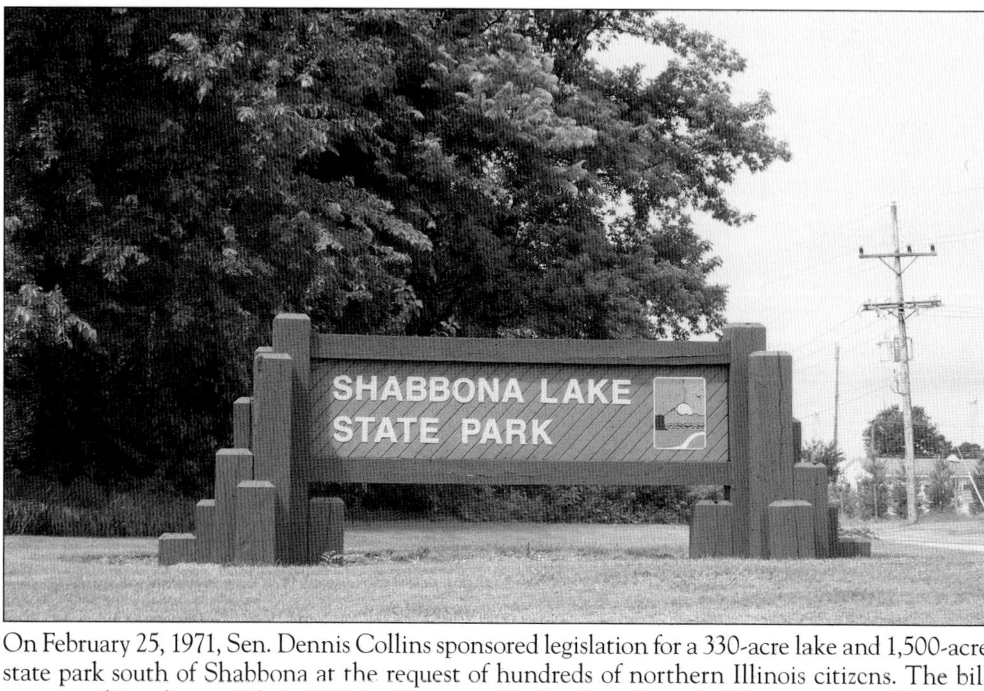

On February 25, 1971, Sen. Dennis Collins sponsored legislation for a 330-acre lake and 1,500-acre state park south of Shabbona at the request of hundreds of northern Illinois citizens. The bill was signed into law on July 3, 1971, by Governor Ogilvie, but the park did not open until January 1, 1978. Founding fathers were Howard Mullins, Marvin Colby, Milo George, Floyd Challand, Porter Martin, and Fritz Vandre.

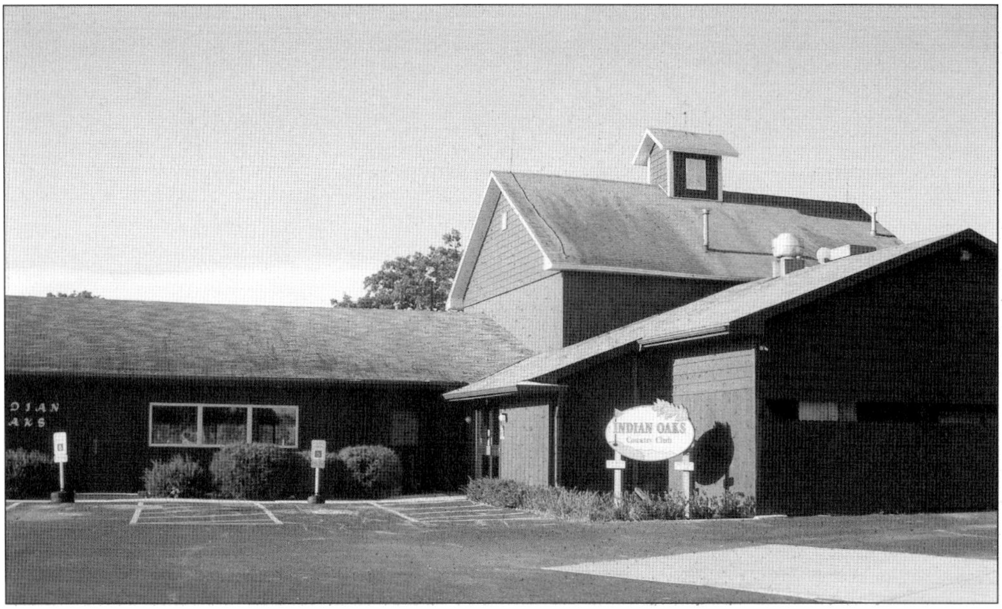

Indian Oaks Country Club is a nine-hole, 3,475-yard golf course. It opened for play on June 19, 1965, and is located southeast of Shabbona on an 80-acre tract at the corner of Indian and Preserve Roads. In the spring of 1965, an old barn on the premises was converted into a clubhouse and pro shop, including a bar and restrooms.

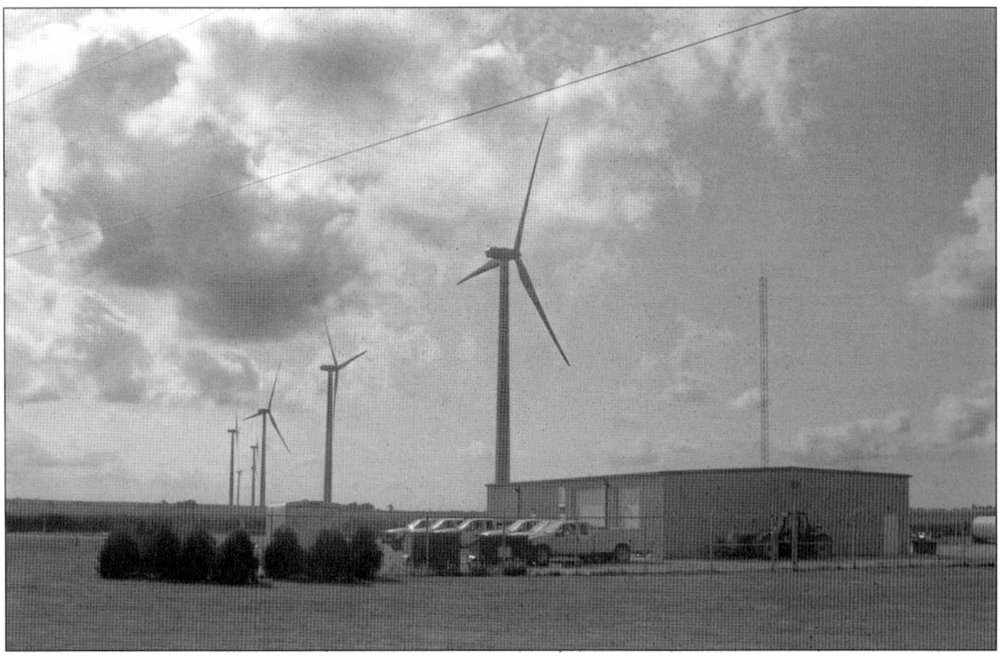

Florida Power and Light brought wind power to DeKalb and Lee Counties in 2009. Each wind turbine is 262 feet tall from the ground to the center of the blade hub. A total of 145 General Electric turbines are capable of generating enough electricity to power more than 54,000 homes. The Wind and Energy Center is operated by NextEra Energy Resources and has a building on the west side of Shabbona.

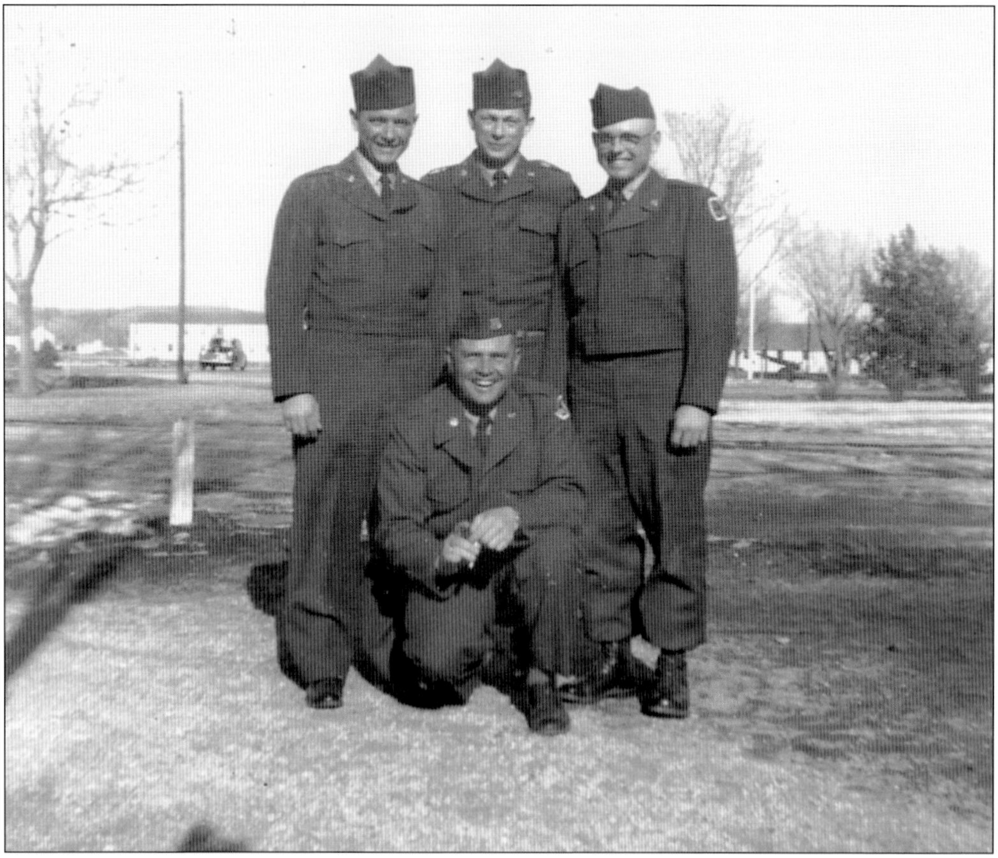

This photograph shows servicemen from Shabbona. They were stationed in Germany as part of the occupation forces during the Korean War, which lasted from 1950 to 1953. Those pictured are, from left to right, (standing) Bob and Jim Barr and Dale Herrmann; and (kneeling) Vernon Herrmann from Shabbona. (Courtesy of Pat Herrmann.)

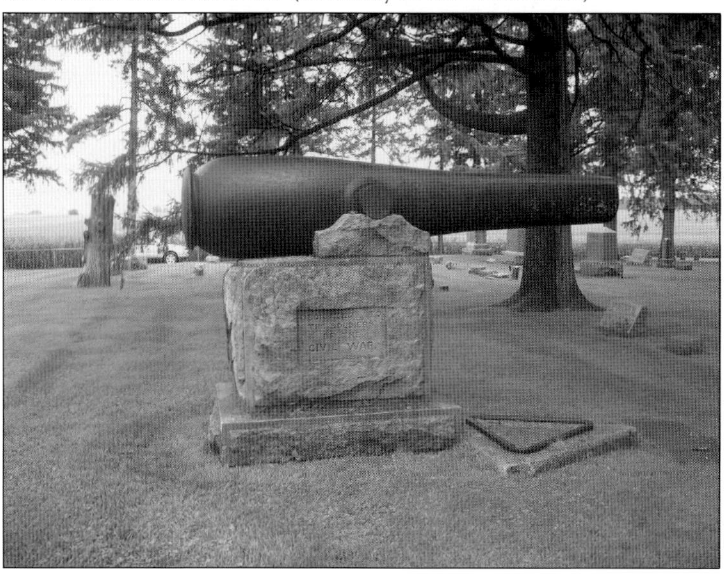

This cannon is located at Johnson Grove Cemetery on Shabbona Grove Road, southeast of Shabbona. It is an eight-inch Columbiad 12-powder M1857 Civil War cannon. It was erected in 1905 and reads, "Dedicated to the Soldiers of the Civil War." There is also an area dedicated in the cemetery for veterans to be buried free of cost.

Ten

SERVING OTHERS

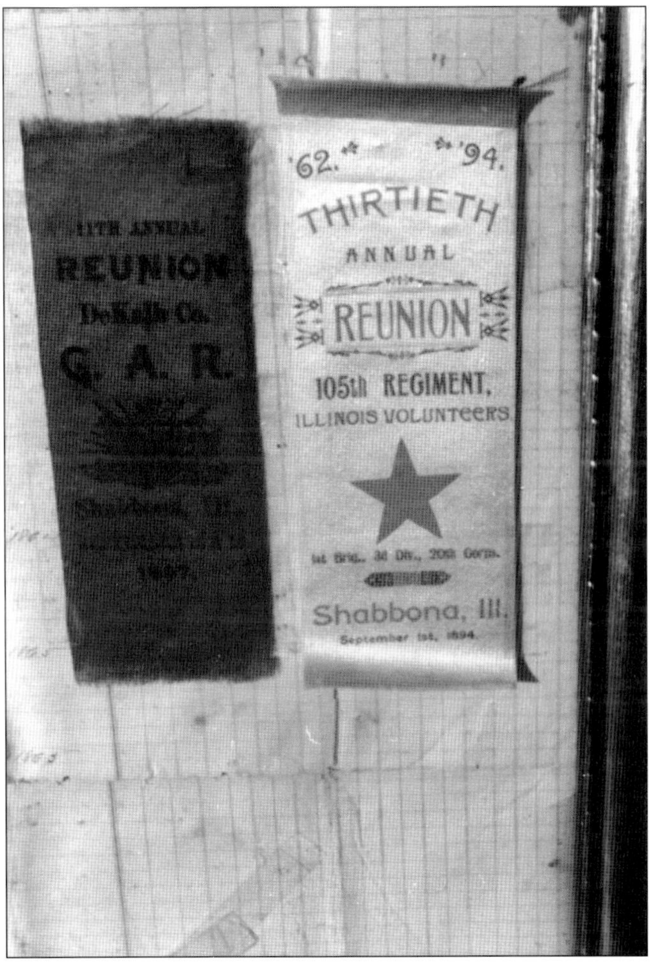

The 105th Regiment, Illinois Infantry Volunteers, was mustered into service on September 2, 1862, at Dixon. From there, the volunteers went to Chicago, where they received training. Many served eight months before returning home. Others were killed in action. The banner is from the 105th Volunteers' 30th annual reunion held in Shabbona on September 1, 1897. Rev. J.M. Green responded to the address of welcome, and the old soldiers made speeches.

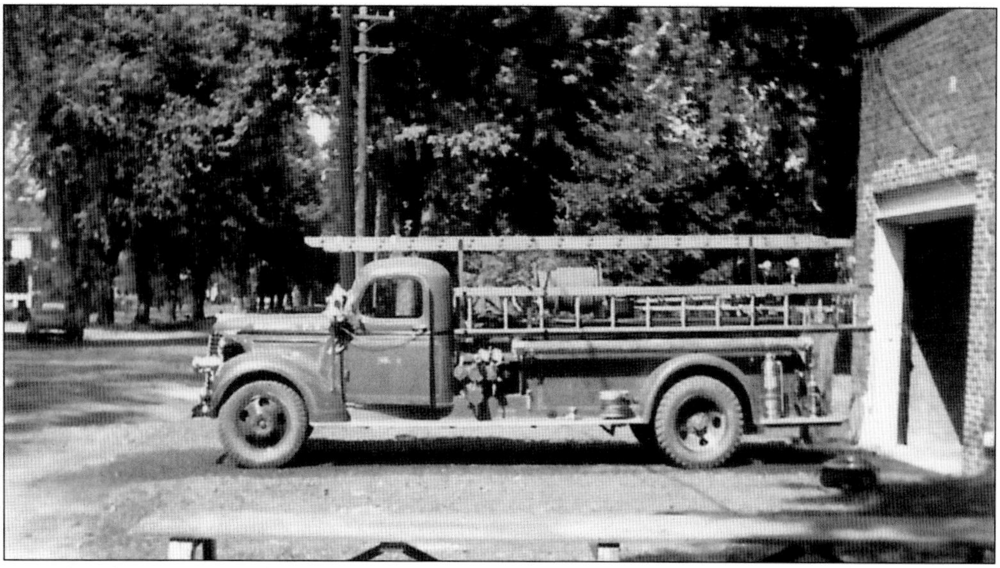

In 1930, a fire protection district was established by all property owners in the district each paying a portion to support it. Volunteers were organized in 1938. Later, a truck was purchased, but it only carried 250 gallons of water. It was suggested to farmers to bring a couple milk cans filled with water if they were going to the fire. Over the years, several pieces of fire equipment have been purchased. In March 1973, a new ambulance was purchased. The picture above is of Shabbona's fire station on Navaho Street. The picture below was taken in 1994, when Tracey Falk, standing fifth from the right in the middle row, was fire chief and Fritz Vandre, sitting second from the left, was mayor of Shabbona. Frank Ottengheime, standing fourth from the right (middle row), is the current fire chief.

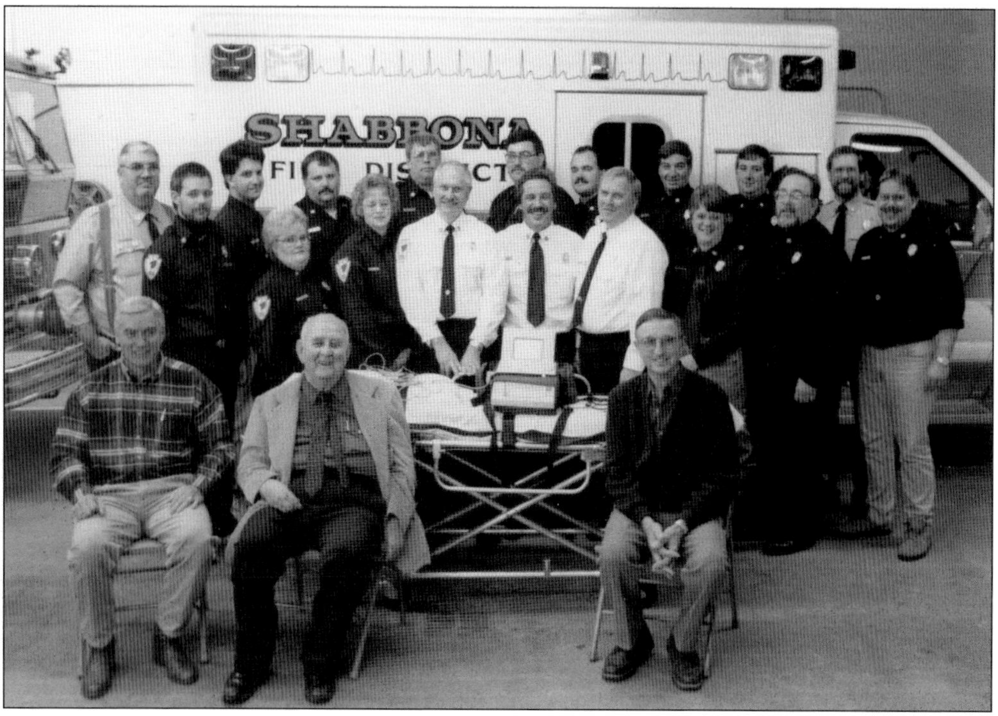

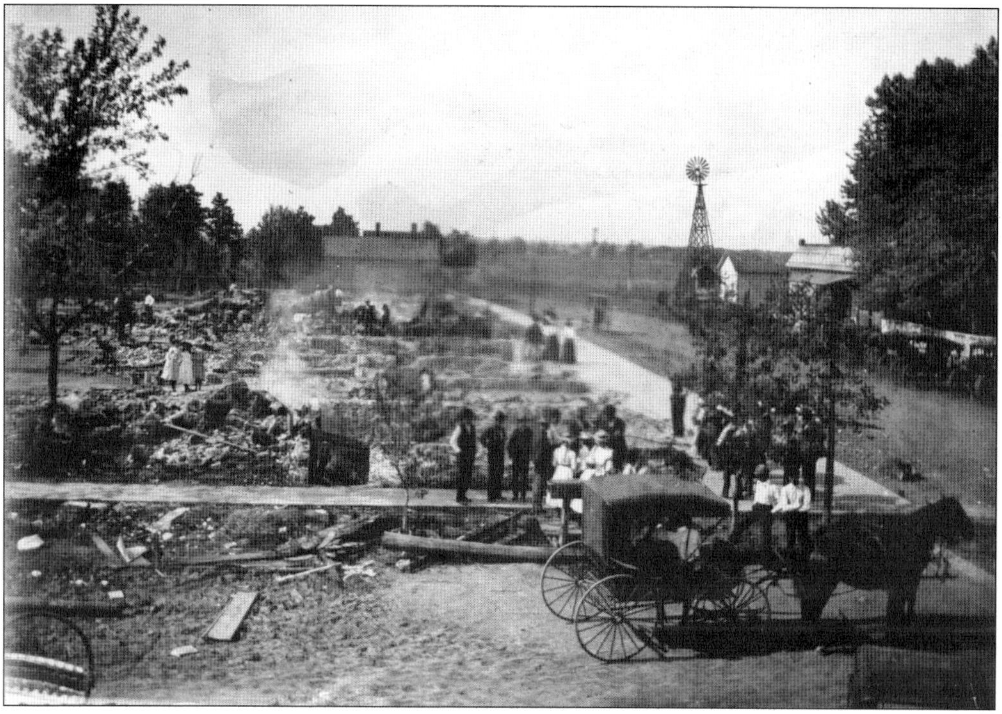

In 1914, a large number of buildings in the town of Lee burned, including the blacksmith shop, the light plant, and the implement sales. By the time the alarm was given, there was little that could be done to save the buildings and their contents. The farmers of the community and the businessmen rallied to help, both with money and their labor. As a result, the village decided they needed better protection, and they decided to dig a deep well with a large storage tank, laying water mains at strategic points. Also, the wooden buildings were replaced with brick structures. Old Lee County records show Henry Eide as chief in 1914. Lee has a fire department building on the south side of town.

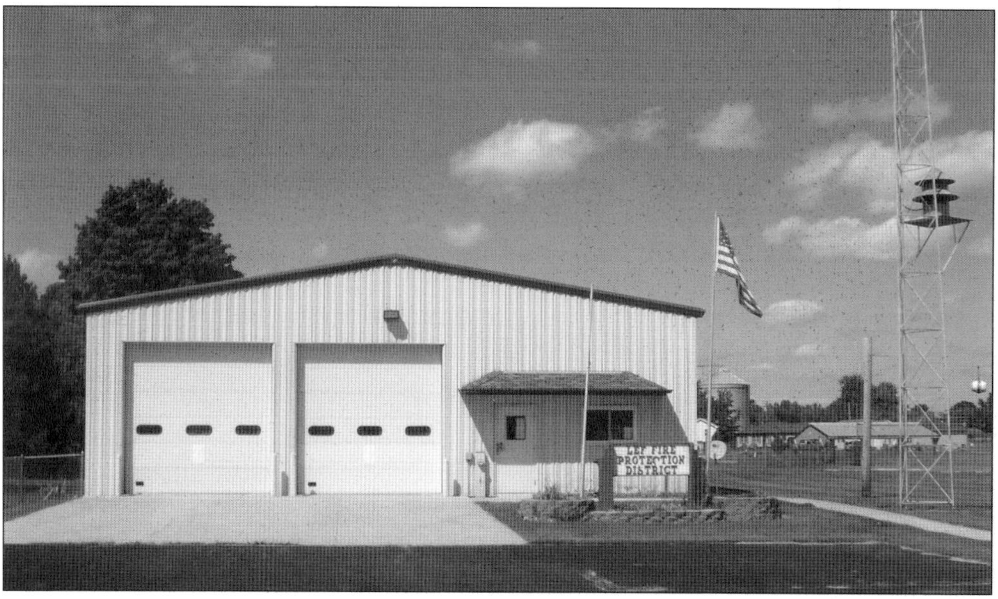

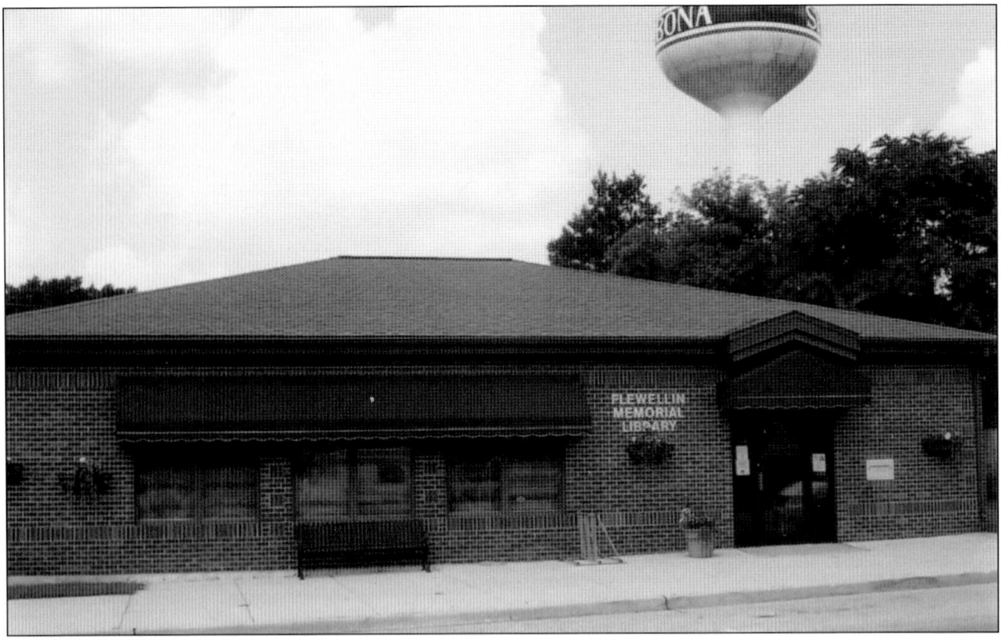

In the late 1880s, Shabbona's library occupied a part of the lecture room at the Congregational Church. The Shabbona Woman's Club organized the library in 1939. On April 20, 1943, the library became a village library supported by taxation. A library was built in 1990 with money donated by Mabel Flewellin. In 2001, an addition was added. The Flewellin Memorial Library is located on Comanche Street in Shabbona.

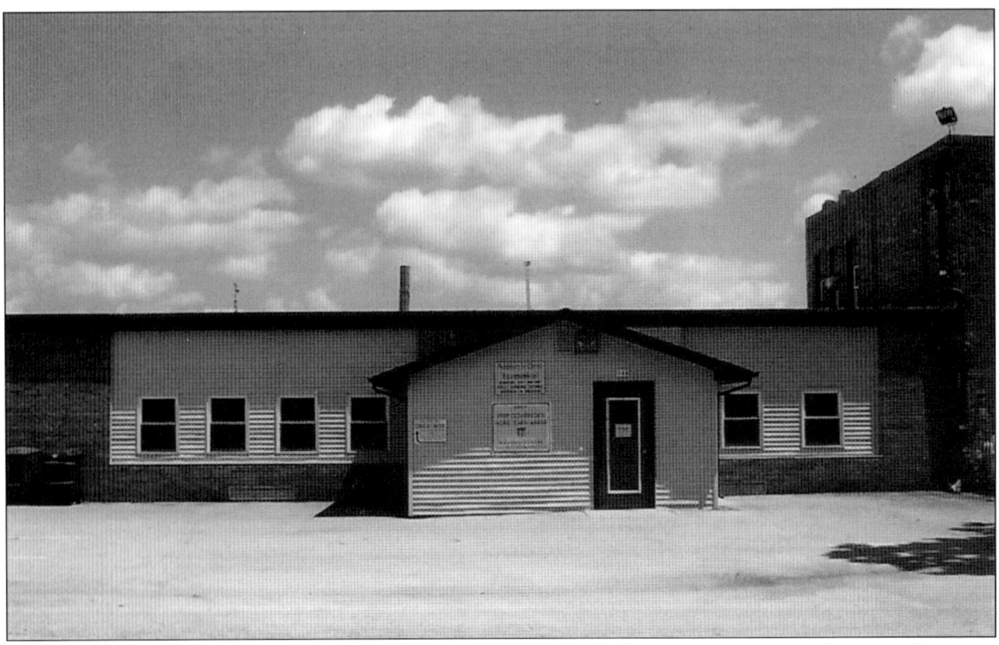

About 100 volunteers from eight area churches and five townships came together from neighboring communities to operate the Neighbors in Christ Ecumenical (NICE) Food and Clothing Center. For 10 years, the organization was in Shabbona, and in 1999, members were able to purchase their own building in Lee.

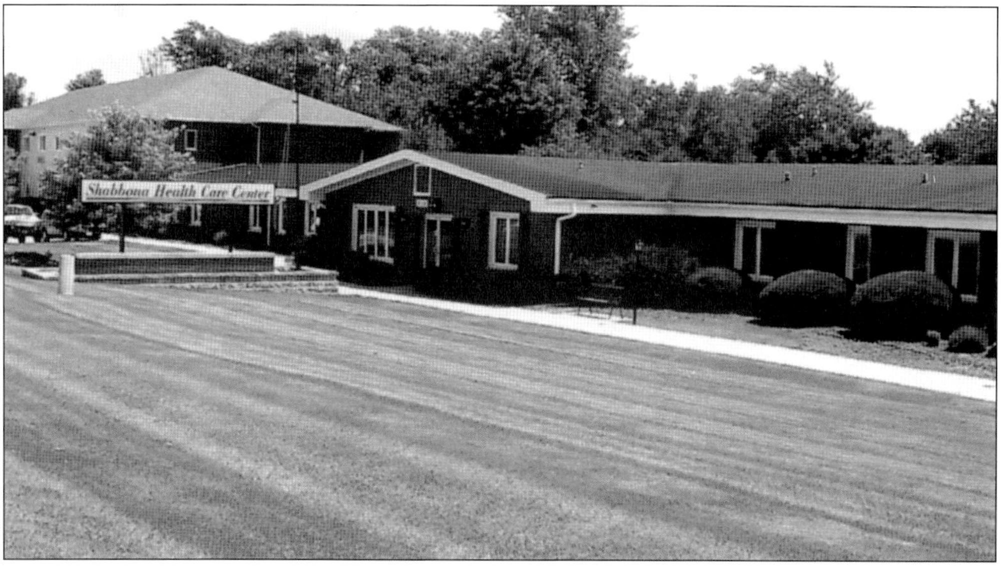

At 409 West Comanche Street in Shabbona, Shabbona Healthcare Center opened in June 1970 with 42 beds. An enlarged addition was completed in February 1972. National Care and Convalescent Industries, Inc., with Alvin Martin of Yorkville, Illinois, as its manager, built and operated the nursing home for many years. It is now under new ownership. Shabbona Assisted Living opened its doors, at 407 West Comanche Street in 2006, to help residents with housing. It is located next door to the nursing home.

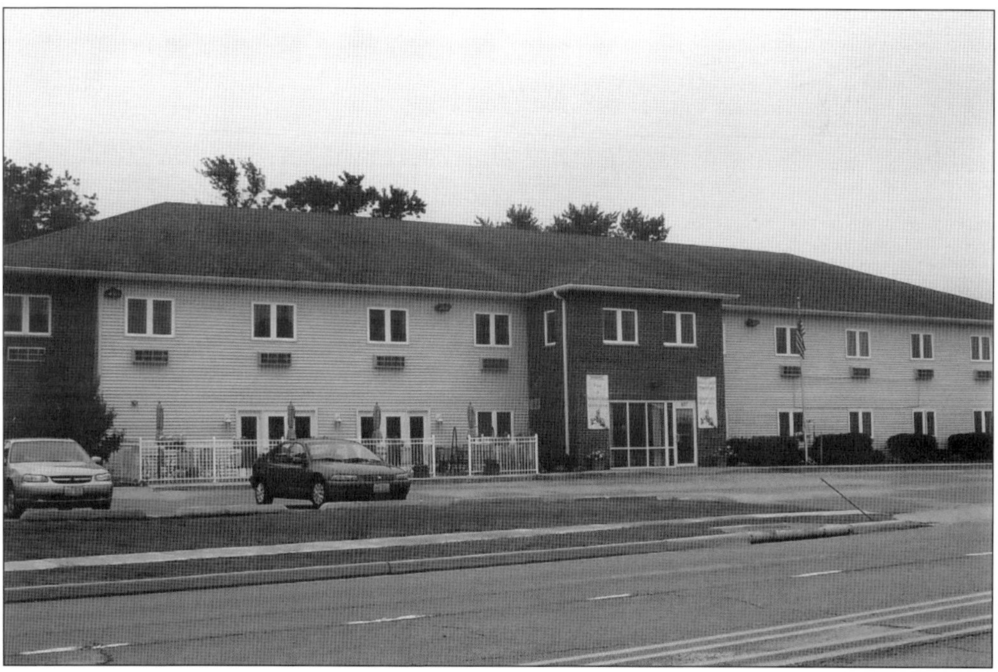

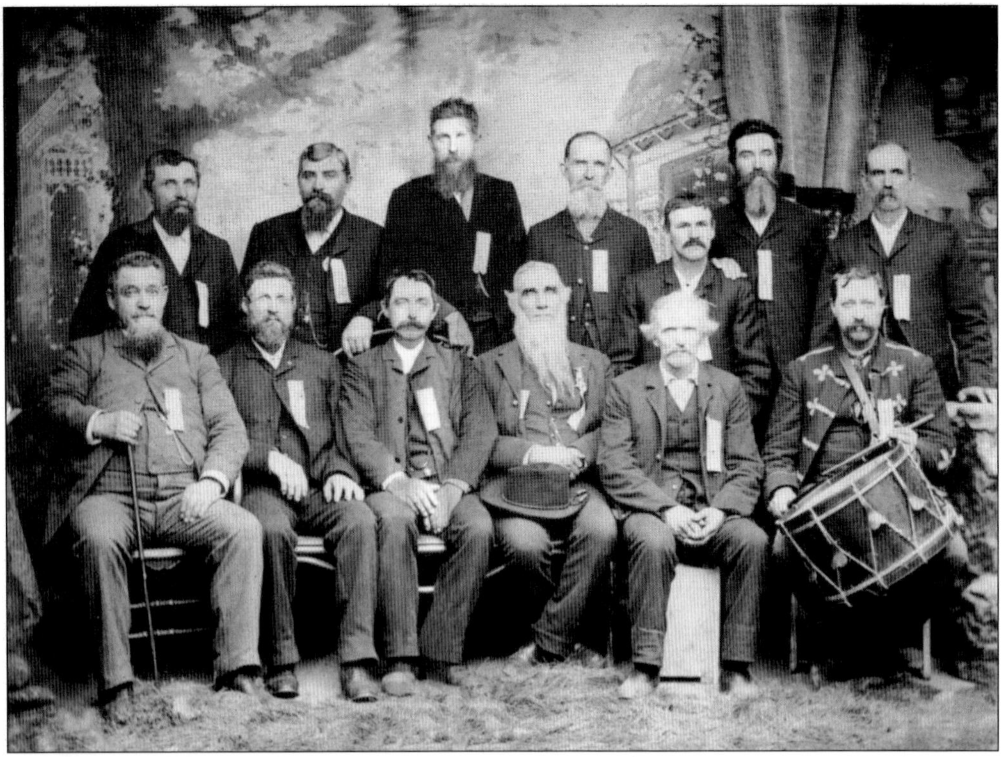

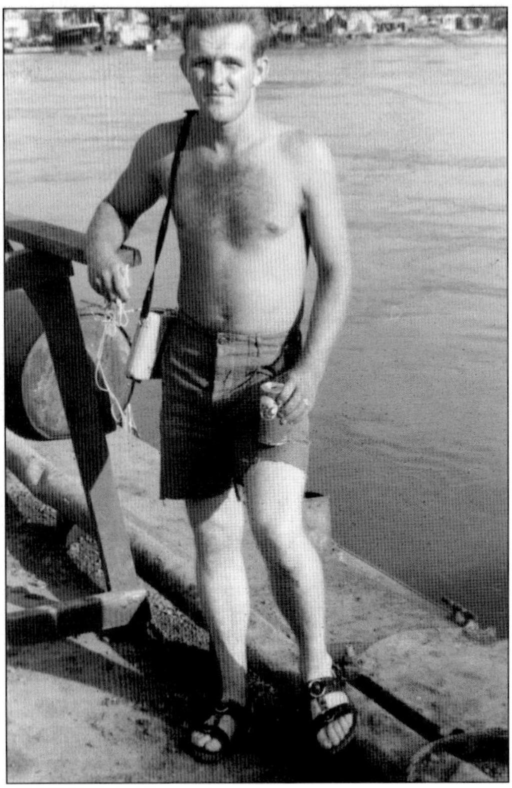

The 105th Illinois Infantry (Grand Army of the Republic) was organized at Dixon, Illinois, and mustered in for three years of service on September 2, 1862, under the command of Col. Daniel Dustin. It was recruited in DeKalb and DuPage Counties. Some men identified in the picture from the area are John Mullins, A.J. Parter, John McCormick, Tom McCormick, W.H.O. Stevens, Captain Allen, and Hiram Morry. (Courtesy of Ella Shrader.)

This picture of Robert Carlson was taken in 1969 at the Mekong Delta, Vietnam. Acting staff sergeant Carlson and his group were in charge of building artillery pads and helicopter hangars and airstrips. Before going to Vietnam, he was in Germany for two years, having done basic training at Fort Campbell, Kentucky. Carlson is a current member of the Shabbona-Lee-Rollo Historical Museum Board. (Courtesy of Robert Carlson.)

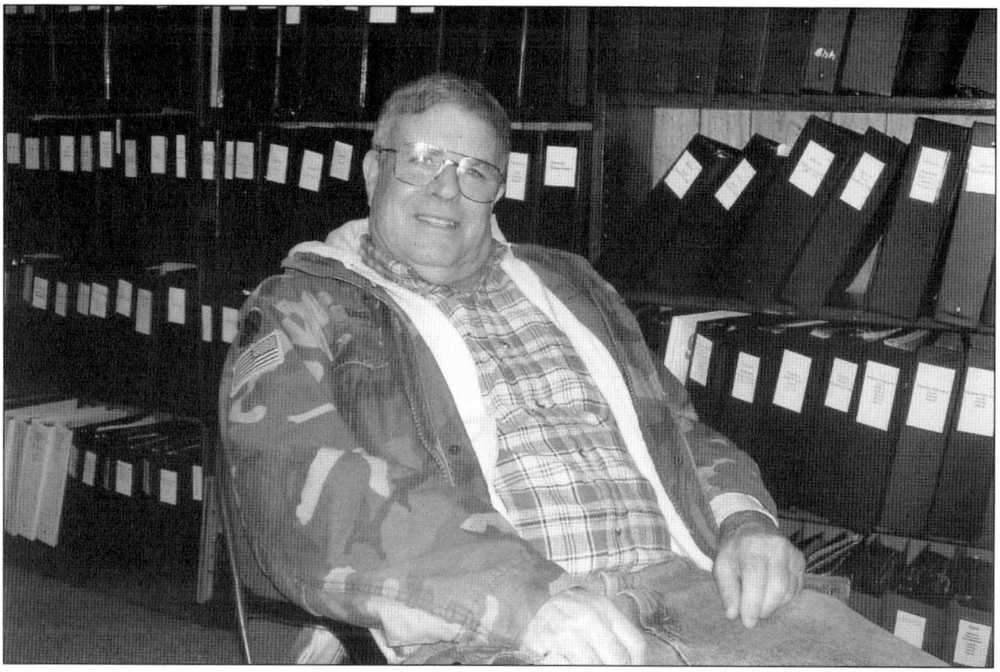

Jim Cox, current member of the Shabbona-Lee-Rollo Historical Museum Board, was part of Operation Iraqi Freedom. While a member of the Illinois National Guard, he was sent to the Middle East at age 58. He also served three years in the Army when just out of high school. Jim has since retired. Clark DeBolt (below), a member of the World War II Fliers Group of the Navy, served from 1943 to 1946. He qualified for either submarine or aviation duty. He chose to be part of a 13-man crew doing aviation duty, being an aerial gunner. When a pilot was down, they would do open-sea landings and pick him up.

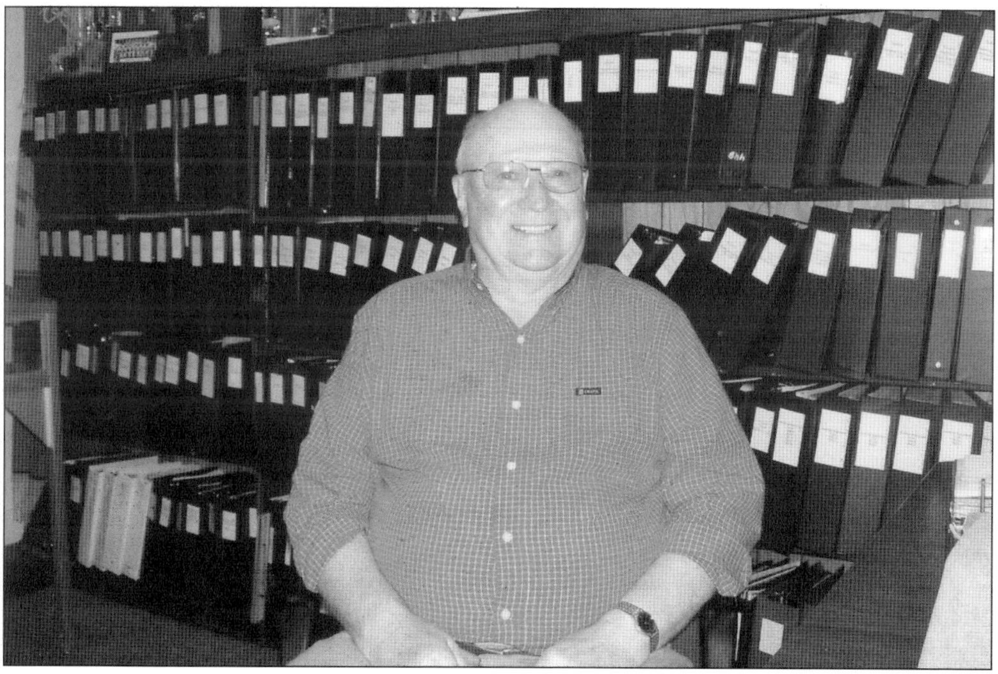

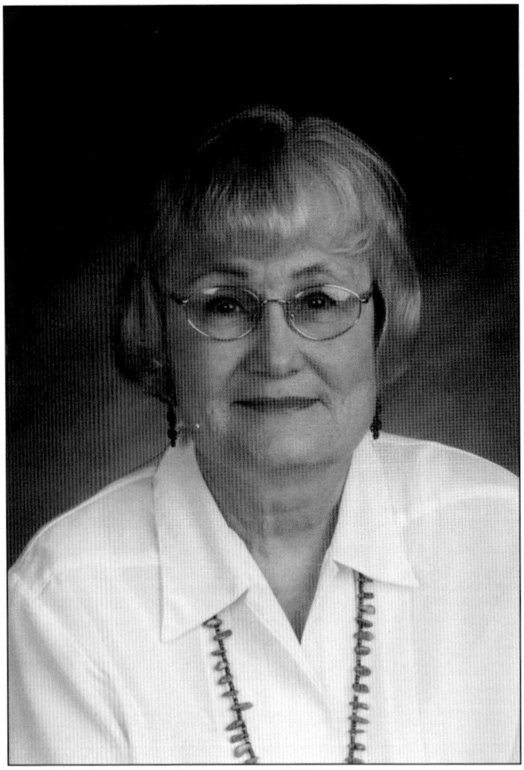

The president of the village board in Shabbona is Claudia Hicks. She was born in Great Falls, Montana, where she grew up and attended the state university. She is married and has three children. She was secretary at Shabbona High School for a number of years before retiring. Since 1972, Hicks has been associated with the village board, first as a clerk and for the past eight years as mayor. The president of Lee Village Board is Richard Voris. He has two business degrees from Northern Illinois University and served as business manager for several years for a public school district. Currently retired, he and his wife have lived in Lee for four and a half years, and he has been on the village board since May 2009.

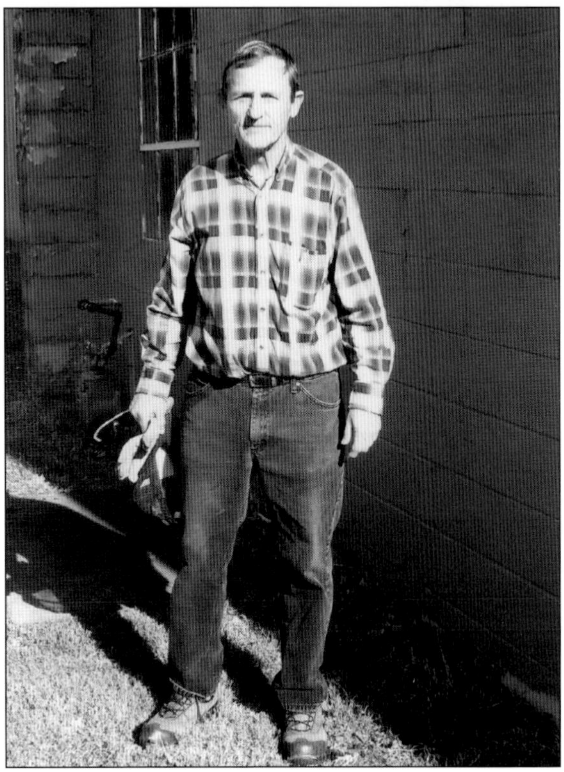

The following is from the Shabbona Centennial Committee (1872–1972) to the community: "Our goals have been to pay Honor to our past, look at the present, think of our future and join in with our neighbors and friends to work toward an end that would be not only fun and enlightening to all of us in the knowledge of our history but mainly rewarding in the fellowship coming from working together."

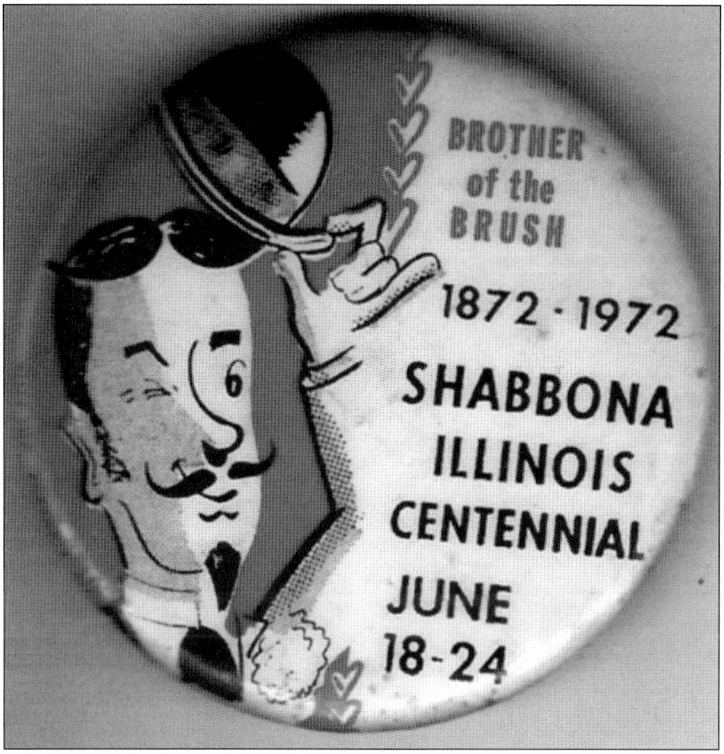

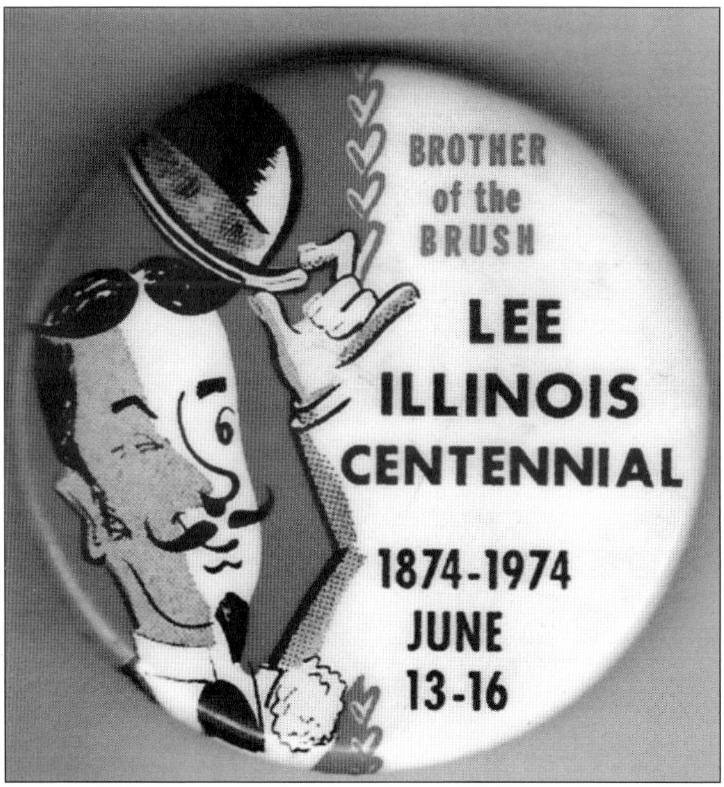

The following is from the Lee Centennial Committee (1874–1974) to the community: "We have looked into the past and find that we do have a heritage of which we can be very proud. As we turn to the future, let us always keep in mind that one fact that has been brought out by our celebration—by cooperating and working together, we can accomplish the seemingly impossible."